Walter Gibbons Cox

Artesian Wells as a Means of Water Supply

Including an Account of the Rise, Progress, and Present...

Walter Gibbons Cox

Artesian Wells as a Means of Water Supply
Including an Account of the Rise, Progress, and Present...

ISBN/EAN: 9783744750073

Printed in Europe, USA, Canada, Australia, Japan

Cover: Foto ©berggeist007 / pixelio.de

More available books at **www.hansebooks.com**

ARTESIAN WELLS

As a Means of Water Supply,

INCLUDING AN ACCOUNT OF

THE RISE, PROGRESS, AND PRESENT STATE OF THE ART OF BORING FOR WATER IN EUROPE, ASIA AND AMERICA;

PROGRESS IN THE AUSTRALIAN COLONIES;

A TREATISE ON THE WATER-BEARING ROCKS; PERMANENCE OF SUPPLIES;

THE MOST APPROVED MACHINERY FOR, AND THE COST OF BORING, WITH FORM OF CONTRACT, ETC.;

TREATISE ON IRRIGATION FROM ARTESIAN WELLS;

AND ON

SUB-ARTESIAN OR SHALLOW BORING, ETC.

BY

WALTER GIBBONS COX, C.E.,

ASSOC. INST. C.E.'S, LONDON, 1869.

PRICE, SEVEN SHILLINGS AND SIXPENCE.

BRISBANE, SYDNEY, MELBOURNE, AND ADELAIDE.

PUBLISHED BY

SAPSFORD & CO., ADELAIDE STREET, BRISBANE.

Sep., *1895.*   1895.

D. VAN NOSTRAND COMPANY,

NEW YORK.

DEDICATED

BY PERMISSION

TO MY ESTEEMED AND EARLIEST PATRON

IN QUEENSLAND,

Sir THOMAS M'ILWRAITH, C.E., K.C.M.G.,

LATE PREMIER OF QUEENSLAND.

PREFACE.

A FIRST edition of this Book entitled, "On Artesian Wells as a Means of Water Supply for Country Districts," was published in Brisbane in 1883. It was based upon papers published for me in the *Leader* newspaper, Melbourne, in 1878, one year before the first Artesian Bore was made in Australia—that at Kallara Station, New South Wales, in 1879, and two years before what is claimed to be the first artesian bore made in Australia—that at Sale, in Victoria—was commenced, advocating the utilisation of artesian supplies in this country.

A series of Papers on "Artesian Wells and Well-boring" was also published for me in *The Brisbane Courier* in 1883, four years before the first artesian boring was begun in Queensland—that at Blackall by the Government—and five and a-half years before the first successful bore—that at Barcaldine by the Government—was started.

A great deal of further information appears in this second edition, covering—it is believed as fully as can be in a work of its dimensions—all matters pertaining to artesian water-supply to the present date, and an account of the progress made in all the Australian colonies.

In the preparation I have consulted the works of the following authorities:—

TRANSACTIONS INSTITUTION CIVIL ENGINEERS, London.
ANDRE ON MINING.
M. DEGOUSEE, Artesian Wells, Paris, 1847.
M. ARAGO, Essay on Artesian Wells.
M. TOURNEL, C.E., Algeria, 1885.
REVUE SCIENTIFIQUE. Article, "Algeria," 1886.
OFFICIAL REPORT to the Governor-General of Algeria on Artesian Wells in the Province of Constantine, 1860-1864.

REPORT OF AGRICULTURAL DEPARTMENT, Washington, on Artesian Wells in the United States, 1889-90.
PROFESSOR C. W. HALL, Minnesota, U.S.A.
C. N. HEWITT, Minnesota, U.S.A.
R. J. HINTON and COL. C. T. NETTLETON, C.E.'s Artesian Underflow Department, Washington, U.S.A.
PROFESSOR T. C. CHAMBERLAIN, Wisconsin, U.S.A.
F. T. B. COFFIN, State Engineer, Dakota, U.S.A.
INDIAN REVENUE REPORTS—On Irrigation in the Madras Presidency.
TRANSACTIONS ROYAL SOCIETY, New South Wales.
REPORTS OF WATER SUPPLY DEPARTMENTS of Victoria, New South Wales, South Australia and Queensland.
H. C. RUSSELL, B.A., F.R.S., &c., Government Astronomer, Sydney.
E. F. PITTMAN, F.G.S., Government Geologist, Sydney.
PROFESSOR J. P. A. STUART, M.D., Sydney.
PROFESSOR A. LIVERSIDGE, Sydney.
PROFESSOR J. W. E. DAVID, B.A., F.G.S., Sydney.
J. W. BOULTBEE, Water Conservation Department, Sydney.
A. B. MONCRIEFF, M.I.C.E., Engineer-in-Chief, Adelaide.
R. L. JACK, F.G.S., F.R.G.S., Government Geologist, Brisbane.
J. B. HENDERSON, M.I.C.E., Government Hydraulic Engineer, Brisbane.
Also the Government Reports of all the Australian Colonies.

I have also consulted the works of other able writers on the subject, too numerous to particularise, and for progress of boring operations the "Country Correspondence" of the public press, from which particulars of the private bores made to date have been mainly compiled; but I may express my grateful special acknowledgments to A. B. Moncrieff, Engineer-in-Chief, South Australia; H. C. Russell, Government Astronomer, New South Wales; C. F. Pittman, Government Geologist, New South Wales; Professor J. P. Anderson Stuart, New South Wales; H. W. Meakin, Secretary Water Supply Department, Victoria; Professor C. W. Hall, Minnesota, U.S.A.; R. L. Jack, Government Geologist, Queensland, all of whom responded in a most courteous and liberal manner to my application for their views, writings and reports.

FRONTISPIECE.

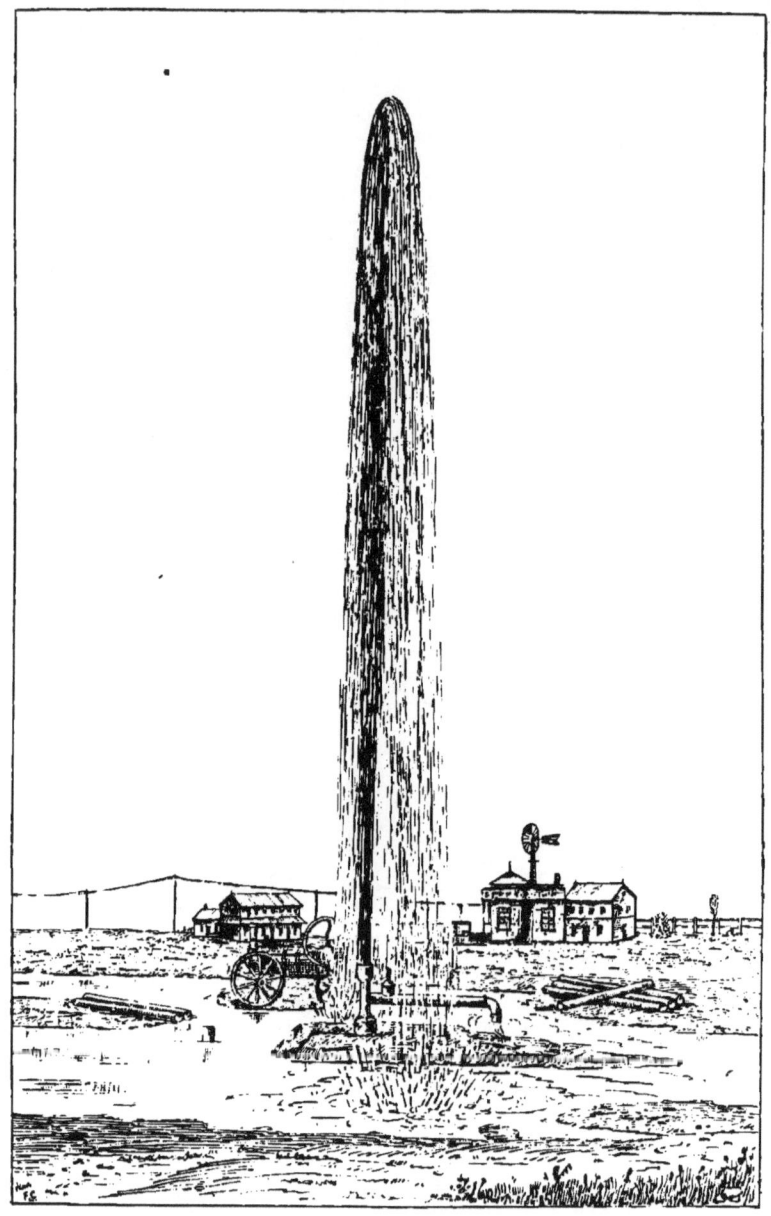

ARTESIAN WELL AT WOONSOCKET, SOUTH DAKOTA, U.S.
Throwing a four-inch stream 61 feet high; depth, 725 feet; bore, 6 inches; discharge, 8,000 gallons per Minute, or 11,520,000 *gallons per diem.*

ARTESIAN WELLS

As a Means of Water Supply.

INTRODUCTION.

THE story of the rise, progress and development of the art of well making is a long but exceedingly interesting one, and in no country in the world is there likely to be, in the near future, so great a degree of interest taken in it, nor such valuable results accrue from the development of it as in Australia.

Although a certain number of far-seeing colonists—few in comparison with the great number who were directly interested in the vital question of an increased supply of water to the country districts—had persistently advocated, through a tedious current of years, the necessity of obtaining artesian water by means of boring, it is only comparatively recently that the attention of the Governments, pastoralists and the public has been really attracted to the subject. It had been urged here for many years in the best possible channel of publicity—the public press—by engineers who had devoted their skill and energies to the adoption and development of it, and this during a time of unequalled prolific expenditure in the construction of expensive railways and other public works, and in the development of goldfields, many of which have not as yet paid their way. It is now, unfortunately, only too easy to calculate the enormous saving in live stock alone, which would have accrued if a small portion only of the millions of money that have been expended in the years gone by upon public works had been diverted to tapping the artesian water supplies then undoubtedly at our disposal.

That highly practical and beneficial results had been all along possible was shown by the experience of other countries, and also repeatedly by borings in various parts of the country in which good serviceable sub-artesian water had been obtained at comparatively moderate depths, but when the news came from the first successful (Government) bore, that at Barcaldine, in that normally droughty interior of Queensland, that so large a supply of water was tapped there, and that it overflowed the surface,

B

the utmost resultant confidence was placed in the capabilities of the well-borers, their desirableness was much enhanced, and they became, in fact, quite suddenly famous, reflecting a luminous light upon the departments of water supply which have since, in a very energetic and highly successful manner, prosecuted the search.

To this vast country nature has been exceptionally prodigal with her riches of the vegetable and mineral kingdoms, and climates more or less adapted to every form of life. She has at the same time stinted and sorely tried us in the supply of available surface water. The scarcity of river supply—due to the peculiar nature of the soil, a greater part of the supply passing away out of sight—left us to work out the problem of utilising, by intercepting and distributing artificially, whatever the heavens do send us as it flows, either on the surface or below, or both, to its final lowest level—the ocean. The primitive and very obvious mode of marking the course of flood-water in creeks, gullies and low-lying lands, damming it back and thus conserving it in times of succeeding droughts, had been universally practised in Australia. This principle of conserving water by means of dams for the purpose of stock had been carried out almost exclusively until within a few years ago. Having been adopted in the beginning it had been adhered to, and had entered into the ordinary practice of the improvers of land to the exclusion mainly of other modes of water supply, and large sums of money have been expended, not only in the first construction, but in subsequent repairs and maintenance of these works—if they were not indeed in many cases totally abandoned. The practice had in fact become a national *habit*. With the condition of a large natural surface channel or river-way, as shown by the great Irrigation Works of India, and those recently carried out in Victoria and South Australia, water may be dammed back and utilised on a large scale, not only for general purposes, but for filling the place artificially (by irrigation) of the variable and uncertain rainfall we are subjected to. Where these large natural surface rivers or channels do not exist as a basis for operating, another and most important source of supply which is left to us must be resorted to namely, wells.

Section I.

ANTIQUITY OF BORING.

Of the origin and earliest state of this art, as that of the architecture of the Ancients, no records are left. History is silent on *that* head. The engineer of the present day enquires in vain amongst the records of kings and dynasties for the mode in which those ponderous structures—the Pyramids, for instance—have been wrought in the hardest granite without the aid of iron and moved immense distances over a desert of sand and to enormous heights. It is probable, however, that most, if not all, of the devices of the Ancients for *raising water* have been uninterruptedly handed down to us.

In Asia, the cradle of our race, the former inhabitants had, from the peculiar physical formation of their country and its climate, the strongest possible inducement—that of absolute necessity—to cultivate the art of raising water, and we have it on record in Holy Writ that the works constructed for this purpose were both extensive and numerous, and the great value placed upon the inestimable boon of "living water" is evident from the phrases alluding to it running through the Scriptures. At first no more was accomplished than digging shallow cavities in moist places, the depth being increased so as to contain the surplus water which might flow into them at certain intervals. When, however, the discovery of the metals was made—the "iron age" set in—the depth was no longer limited by the rocky formation of the ground, and in the East some of the oldest wells known are dug almost entirely through rock, and to what, until comparatively recently, were considered great depths. The mode of sinking was by hand, and some of these wells are of large diameter. Joseph's well at Cairo, one of the most remarkable wells ever made, is in form an oblong square twenty-four feet long and eighteen feet wide. It was made through solid rock for 165 feet, at which point it is enlarged into a capacious chamber, on the bottom of which a basin or reservoir is formed to receive the water from below. On one side of this reservoir another shaft is continued 130 feet lower, where it emerges through the rock into the water-bearing gravel underneath. The lower shaft is fifteen feet long by nine feet in width. The water is raised by means of a chain of buckets on endless ropes led over a pulley worked by cog-gearing and a capstan bar, to which oxen or horses are attached. On the top of the lower depth the vestibule is connected with the surface by a spiral staircase cut in the solid rock. In this vestibule oxen or horses work the machinery of the lower depth, and raise the water to the reservoir. Although the Temple of Solomon, with his golden house, ivory palaces, and gorgeous gardens are wholly gone to decay and obliterated, the "Pools of Solomon" are, like this well of Joseph's, in a high state of preservation, and dispensing the "heavenly fluid" as of old. Ephesus, too, is no more; and the Temple of Diana, which, according to Pliny, was 220 years in building, and upon which was lavished the talent and treasures of the East, the pride of Asia and one of the wonders of the world, has vanished, while the wells and fountains which furnished the city with water, remains as fresh, as perfect, and as "life giving" as ever.

Besides extensive hydraulic works in Ancient Persia and Assyria, Egypt and China, we find traces of equally extensive undertakings in the Western Hemisphere, in pre-Christian America. At the city of Mexico remains still exist of great wells and dykes for the supply of water which were constructed by Montezuma; and the Incas of Peru were also great adepts in the art of water supply. In Ancient Greece wells were very numerous. The inhabitants of Attica were supplied with water from them. The ancient Egyptians irrigated the borders of the desert, above the reach of the Nile, from wells. In India not less than 50,000 wells were counted in one district of Hindostan when taken possession of by the British, several of which were of high antiquity. In China wells are numerous and often of large dimensions, and they are even lined with marble. In Pekin they are very common, and some of them are of considerable depth. In this work, as in many others, the nation which we are apt to despise seems to have taken the lead and constructed the earliest, most numerous and most useful works; and when we talk of artesian wells we are apt to forget that they were known in China probably thousands of years ago. Nearly the whole of the vast Continent of Asia depends for its supply of water on wells; and the resemblance between the steppes of Tartary and the great plains of Australia is sufficiently close to warrant the belief that a system successful in the one case may be equally so in the other, and yet in these colonies artesian water had not been obtained or utilised until a few years ago. The scarcity of water is very great all over Persia, too, and the inhabitants depend entirely on wells for a supply. The Hindoos, says Somerat, believe the digging of wells renders the gods propitious to them. Although there is evidence of extensive irrigation works having been carried out in connection with the supply of water from the great rivers in the most remote ages in Egypt, India, and China, the practice in later, though remote, times has been to sink wells to supply each locality by means of irrigation works in connection with them. This method is pursued to this day by the Hindoos, Japanese, Chinese, Tartars, &c.

It will thus be seen that the practice of sinking into the earth's crust for the supply of water has been universal in those countries where great heat, a scarcity of surface water, and an arid soil prevailed. Taking into consideration the important fact that in all those countries human labour was at command to an almost unlimited extent, and at the very lowest possible cost, and that possessed as they were of a high although unique civilization, it would appear that well-sinking was preferred, not only to supply their domestic requirements but that by means of local systems of irrigation by the water obtained from the wells, the pastoral and agricultural requirements of those countries were supplied. If a system of canals or extensive irrigation works in connection with the great rivers had been preferable they doubtless would have been carried out. The matter of labour or cost does not seem to have greatly concerned the powers of the ancient countries of the world. In those countries the heat during the greater part of the year was excessive; there was a scarcity of internal water supply from numerous rivers; the rainfall was variable and uncertain producing a dry atmosphere and a parched and temporarily a barren soil; the custom was, therefore, to sink into the earth for that treasure which was considered and found to be worth seeking. That treasure, the rain from above, so precious even in its variable and uncertain fall, was known to pass down into the earth leaving but a portion of its volume to nourish the surface behind, as it descended lower and lower until it reached its impervious water-bearing bed conserved in limitless water-bearing strata chasms and reservoirs, and awaiting but the skill, energy, and industry of man on the parched surface above to call it forth.

Greece and Rome, particularly the former, were great well makers; they showed much skill in construction and in such repute was the fluid held that in the embellishment of the public wells in their cities that fine architectural taste was lavished upon which no succeeding generation of any country has been able to improve. It was the Romans who introduced well-making into Britain, and it would be curiously interesting to know what amount of money was expended in sinking, by the costly process of manual labour, the hundreds of thousands of wells which were made there before the era of the Waterworks Companies set in. Some of these Companies in the London district are now dependent upon artesian bores for their main supply.

The mode of sinking wells in ancient times, which seems to have been universal, was by hand curbing, or supporting, the sides in loose ground with brick or wood. In rock the Chinese and Hindoos have used from remote times, or since the commencement of the "iron age" up to the present day, a "trepan" or weight of iron provided with a set of cutting chisels at the bottom of it; this was suspended by a rope, the upper end of which was attached to a long pole acting as a lever, and which was fixed horizontally in a wooden frame. The weight was raised a few feet at a time by the lever, which was then released, and the impact due to the fall of the weight split the rock. It is scarcely necessary to observe that this primitive and inferior apparatus must accomplish its work at the cost, as compared with the superior machinery and materials of the present day, of a great deal of time and labour.

The modern method of boring for artesian water did not come into existence until the beginning of the present century. The French Society for the Encouragement of Agriculture, in 1818, was the first mover in that direction, and since then the improvement has been most rapid. The Chinese still pursue the ancient method of percussion in the boring of artesian wells, and they have followed it, as pointed out, for more than three thousand years. The utilisation of underground water is a leading fact in the history of Oriental regions, and its importance can be realised by an illustration from one country alone. The plateau or high table-land region which forms the greater portion of the peninsula of Arabia is without a single perennial stream or body of surface water. A population of 12,000,000 resides therein; large quantities of wheat, durra, barley, millet, beans and tropical fruits are grown on this high, apparently dry, sterile plateau. Ninety per cent of the water supply which produces the present fertility is drawn from below the surface by means of old bored wells. The striking fact in all of Central Arabia is that of underground water supply. Throughout the eastern part of Oman in the Persian Gulf,* and especially in the villages of Kaseen, as an illustration of the character of these

---

* See Section "Hottest regions on Earth."

supplies, it may be stated that there are 40 wells, the flow from which maintains a population of 30,000 persons. The depth of these underground supplies ranges from 15 to 200 feet.

In British India the system of irrigation by wells is carried on most extensively. Sir James Strachey, in his great work on the Finances of India some years since, placed the area of cultivated land at 200,000,000 acres, of which 28,000,000 acres were cultivated by irrigation. Of this total, about 12,000,000 acres were served by water drawn from wells, and generally by man-power, it being found that the use of the Mot or bullock-well is too costly for that country of cheap human labour. The general verdict among engineers and practical administrative officials in India is that the land irrigated by well-water is more economically served than that irrigated by water from canals. It has been estimated that the area covered in the Indian peninsula by well irrigation is not less than 20,000,000 acres. From accessible statistics of eleven districts, including Madras and the Punjab, allowing 10 acres for each well, 392,593 are reported serving 3,925,930 acres. The British Government since the famine period has given great encouragement and attention to the system of village wells; allowing bounties for their protection and providing careful regulations for their maintenance. They have found it cheaper and wiser to assist in securing a permanent supply of water, so as to prevent famine, than to feed the people after the famine comes. In Sind, Beloochistan, Cashmere, Afghanistan, Persia, Chinese Turkestan, and in Russian Turkestan, the utilisation of underground waters has always played a most important part. A vast system of natural underground conduits exists throughout the regions named, especially in Persia, that carry the drainage waters of the foot-hills regions for long distances, until they debouche upon the less elevated plains below. So enormous is this supply that vast populations have, for many centuries past, been supported from the fields that were quickened by the application of such waters.

Section II.

FRENCH BORES.

The French have always taken a prominent position in well-boring, and to them is due the introduction in Europe of the artesian system of wells, which has since been almost universally adopted, and has proved a most important factor in the economy and development of those countries in which it has been availed of.

Dégousée's great work—"Guide du Soudeur," on artesian well-boring, Paris, 1847—although not of the most recent date, is a most valuable addition to our knowledge, especially of artesian water-bearing rocks and supplies, and the permanence of them.

The term artesian is derived from Artois, in France, in which province the system was inaugurated, and in which there are artesian wells in the old monastries that have been running continuously for centuries. At Aire one has been flowing for over a century to a height of 11 feet above the surface, and another made in the 12th century at the Carthusian Monastery, at Sillers, has been flowing ever since it was sunk.

Two of the most important wells bored in France are those of Grenelle and Passy, both of which were made for the purpose of supplying Paris with water. That at Grenelle was bored to a depth of 1,798 feet of a minimum diameter of about 8 inches. It discharges 800,000 gallons per diem, and the water is remarkably soft and pure. The one at Passy is 1,913 feet in depth, 27½ inches in diameter, and discharges an uninterrupted supply of 3,795,000 gallons per diem. This latter work was entrusted to Kind, a practical German engineer, who had sunk many wells in Germany, Belgium, North of France, Creusot, &c. Other noted French artesian wells are those at Buttes-aux-Cailles, to a depth of 2,900 feet with a diameter of 47 inches, and at the Sugar Refinery, at Paris, to a depth of 1,570 feet, with a diameter of 19 inches.*

The brilliant success of these French wells brought artesian wells into fashion, great numbers being subsequently made in Great Britain and other parts of Europe and America.

The geological formation, from which the artesian water issues forth is that of the chalk, a part of the Upper Cretaceous series of the "Paris" and also the "London basins." This formation will be found fully described in Section IV., "English Bores."

---

* A further description of these wells appears under section "Machinery."

SECTION III.

ALGERIAN BORES.

In Algeria and the great and ancient desert of Sahara, boring for water has been prosecuted from the earliest recorded time, but the apparatus used by the Arabs was of a very primitive kind, and their bores were lined with hollow palm logs with the result that they caved in from time to time the flow ceasing, and what were formerly oases in the desert have disappeared almost entirely. It was left to the French engineers to alter this state of things. In 1856 the French Algerian Government commenced boring operations with the result that in 1885 there were 490 Artesian wells belonging to the Arabs, and 112 belonging to the French Military authorities, all carried out by the Government. The aggregate daily supply from these wells has reached the enormous quantity of over 80,000,000 gallons. The transformation produced by artesian water upon the sandy wastes and hills of Algeria is described by the distinguished French engineer M. Tournel as amazing. Surely science has a right to be proud of this achievement when it is related that in the year 1802, one of the most appalling cases on record of suffering from thirst occurred to a caravan which was destroyed in travelling from Timbuctoo to Talifet when 2,000 human beings with 1,800 camels lost their lives.

The most remarkable example of reclamation by means of artesian well water is found in the desert provinces or departments of Algeria under the French rule. The area, officially given, of French Algeria is 184,465 square miles. The outlying portion is put at 135,000 square miles. In this total of over 329,415 square miles one half belongs to the Sahara or desert. The European population in 1887 was about 250,000; the native and naturalized were 3,328,549, making a total of 3,578,549. Cultivation by the means of flowing well-waters has been sedulously fostered by the French Colonial Government for both political and economic reasons. Such wells as a means of reclamation began systematically to be bored in 1857, the French engineer M. Jus, having demonstrated in 1856 that the desert was endowed with large supplies of underground water. The total number of wells that have been bored since that date in the departments of Algiers, Oran, and Constantine is stated at 13,135. These wells range from 75 to 400 feet in depth, and the low pressure common to the majority of them forces the water over the small bored casings to a distance of about 2 feet above the ground. The waters are then collected in small ditches, which convey them to the vineyards, date trees, and fields of durra, millet, wheat, &c., which comprise the chief products. In all about 12,000,000 acres have been reclaimed in this way. The Government bores are at least one-tenth of the whole number. The total flow of 100 of these wells in the desert south of Constantine is given at 33,016,000 cubic gallons; that is about 120,000 acre feet per year.

Mons: G. Rolland in the *Révue Scientifique* of June, 1886, says in relation to the province of L'Oned Rir:—" Within thirty years the oases (reclaimed by wells) have increased in value fivefold; the condition of the natives has been improved, together with the complete pacification of the southern part of Algiers, and the population has more than doubled. These oases are composed principally of forests of date palms, the shade of which shelters the growth of their grapes. Without irrigation it would be impossible for the earth to yield anything, even the date palms would not produce fruit. The first well was sunk in June, 1856. When the water poured forth it yielded 1,500,000 gallons per diem, and it was named the "Fountain of Peace." In 1885 there were 114 deep artesian wells bored by the French Government in that province. From 492 other wells in the same region the united discharge is one hundred and a half millions of gallons per diem. The Government wells, some of which have been in use for thirty years, have never decreased in their discharge; on the contrary there is a rapid increase in the total water supply. In the department L'Oned Rir these wells have been the means of reclaiming forty-three oases, which support about 520,000 date palms in full bearing, 100,000 other fruit trees, and 140,000 date palms of seven years' growth. The annual value of such production exceeds 500,000 dollars (£120,000) and the entire increase in value has not been less than fivefold. Since the commencement of the artesian waters exploitation, the French Government has organised an extensive plan of colonization, which they are gradually extending southward, taking in areas that have been regarded heretofore as purely desert in character. By so doing they control the nomadic tribes about them, create new oases, and make settlements where nothing but desert waste existed."

The explanation of the Algerian supplies is very simple. About the desert rise mountain ranges, the drainage of these ranges comes down after heavy rains with great force, disappearing in the sandy regions below. Where the water

ALGERIAN IRRIGATION FROM ARTESIAN WELLS.

OUTFLOW 1,560 GALLONS PER MINUTE, OR 2,246,400 GALLONS PER DIEM.

collects in basins at the foot of the mountains, it evaporates under the influence of the summer heat. Disappearing below the sand, these waters give birth to subterranean sheets of water, which according to the place from which they flow, and the hydrostatic pressure to which they are subjected, have more or less force on being tapped and rising to the surface. The utilization of the Algerian underflow offers a most remarkable illustration of the great importance of such waters in their economic use.

According to an official report* the geological formations represented between the Atlas Mountains and the Sahara are Cretaceous, Miocene, Pliocene, and Quarternary. With the exception of the Lower Cretaceous none of the members of the Cretaceous system appear to be thoroughly pervious. The Cretaceous rocks under a large area of the Sahara are covered under too great a depth of Post-cretaceous deposits to be readily accessible to artesian bores. The Miocene system is water-bearing in places and not too deep to be accessible, but by far the greater portion of the artesian water in the Sahara is obtained from Pliocene formation.

The Pliocene strata of the Sahara are formed of white, grey, red, and yellow sand with clay beds and gypsum. The last is in places, as at Bard Ad, in the Oned Rir district, over eighty feet in thickness. Beds of hard limestone over a yard in thickness are met with occasionally. Immediately overlying the water-bearing sands and gravel is a very hard band of rock, a siliceous cement a foot or more in thickness. These Pliocene strata are of Lacustrine origin, and lie in a broad basin, which through earth movements has been thrown into a succession of folds.

Mr. G. Rolland in a paper read before the Geological Society of France, September, 1885, entitled "Ou et comment s'aliment les eaux artésien du bassin de l'Oned Rir," states that these strata derive their supplies of water partly from direct percolation of rain, partly from rivers, especially from those which take their rise in the Atlas Mountains of the North. Part of this water percolates into the permeable soil and finds its way into the deep alluvial deposits which dip towards the interior of the basin. The Pliocene strata are also partly supplied with water by springs rising from the Cretaceous strata, as for example the beautiful springs of Western and Central Zab and those of the Djerid.

---

*Rapport à Monsieur le Gouverneur Général de l'Algérie sur les Forages artésien executés dans la division de Constantin de 1860 a 1864, quoted by Professor David, Trans. Roy. Soc., New South Wales, 1893.

The waters from the Cretaceous springs disappear in the alluvial deposits of the Pliocene formation and form small underground rivers which flow towards the south and become united at a depth to form a main stream, draining towards the south-south-east under the Lacustrine basin of the Pliocene formation. The Cretaceous beds being continuous under the Pliocene basin and dipping towards the Sahara, some of the springs derived from their porous strata break out considerably below the surface, and these of course assist the superficial springs in supplying water to the beds of the Pliocene. The underground streams of the Pliocene basin flow in a number of reticulated underground channels like meshes of a net. These have been distinctly traced by means of bores for a length of one hundred and twenty kilometres (about seventy-four and a half miles).

That the water circulates in distinct channels has been proved not only by the distribution of the successful bores, but also by the fact that small fish, river crabs, and fresh water mollusca are brought up in considerable quantities at some of the bores. (See Section XI., "Queensland.")

Section IV.

ENGLISH BORES.

AMONGST the various strata of the earth in which, or in proximity to which, water is found, the chalk formation affords remarkable instances of successful operations. This formation, itself in the "London basin," and in connection with the "lower greensand" which lies immediately underneath it in France, has produced the water of the artesian wells of London and the south coast of England and the great wells of Paris. The upper cretaceous formations of the British Isles includes the chalk, reposing first on variable deposits of calcareous, siliceous and. argillaceous rock called the upper greensand, and then on a bed of tough clay called the gault, very impermeable to water and very persistent wherever the chalk formation has been reached. The gault generally maintains its essential character and appearance, and is very important in keeping up the water contained in the chalk and preventing it from passing down into the underlying sands of the lower cretaceous series. The general aspect of chalk varies with its condition, moisture, and the degree of exposure it has undergone. It has been found from recent experiments that this rock is capable of receiving into its mass a quantity of water amounting to more than five gallons for every cubic foot of rock beyond the quantity usually contained in dry chalk under ordinary exposure. The uniformity of chalk as a rock formation is one of its most remarkable characteristics. It admits everywhere of the percolation of water, receiving into itself and conveying to its lower bed the water that falls on its surface. It has been found that each square mile of dry upper chalk one yard in thickness contains nearly *three millions and a-half gallons* of water, but the same quantity of rock is capable of absorbing, and it would contain if saturated, upwards of *two hundred millions of gallons*. The surface of "permanent wetness," dependent upon an average rainfall, is so far above the lower surface of saturation as to ensure a supply at least equal to one-half the rain falling in the whole district. This chalk water-bearing stratum has a superficial area in England of 3794 square miles, upon which the rainfall is nearly equal to four thousand millions of gallons daily, or equal to five times the summer stream of the Thames, and as the surface is almost universally pervious—scarcely any rain running off even during the heaviest storms—it follows that there is an inexhaustible

supply of water to meet every possible requirement, the well at
Bushy Meadows alone yielding 1,800,000 gallons daily. It has
been found that by far the greatest portion of the rain which
falls upon the chalk formation, encountering no impervious bed,
continues to descend through various fissures until, arrested by
the bed of gault clay lying beneath the chalk, it fills the lower
cavities, and accumulates to such a height as to force its way
through subterranean passages connecting with the sea. In this
manner an enormous amount of this water is discharged into the
shingle or sand which covers the coast, and even into the bed of
the sea itself. It has been found that in an area of 1200 square
miles in the chalk district, with a rainfall of 20in. per annum,
one-half (or 10in.) has reached the lower fissures. This depth
of 10in. of rain per annum is sufficient to supply the immense
quantity of 400,000,000 gallons per day for every day of the
year, which at present finds a vent and is discharged along the
coast.

This "London basin" has been utilised by boring to an
enormous extent, the wells at Chatfield and Arnwell yielding
over 4,000,000 gallons per diem. The New River Waterworks,
the Houses of Parliament, Bank of England, the fountains in
Trafalgar Square, and many of the prisons are supplied from
this source. The London Jute Company possess a very large
artesian supply at Ponders End, and the Kentish Town Waterworks derive their supply from the same source. On the
southern and eastern coasts, extending as far north as Bourne,
in Lincolnshire, artesian wells exist. The well at Bourne was
bored through Oolitic strata to a depth of 95ft. Below the
alluvial a limestone formation 32ft. thick was met with and
continued until a stratum of hard rock, 6ft. thick, was passed and
the water-bearing stratum was reached, and a supply of 567,000
gallons per diem, rising to a height of 40ft. above the surface,
was obtained. At Manchester, in the north of England, twenty
artesian flowing wells have been drilled through red sandstone
of diameters from 13in. to 18in., at depths from 145ft. to 466ft.,
supplying from 34,000 to 806,000 gallons per diem.

DEEPEST BORES IN THE WORLD.

PRUSSIA is to be credited with the deepest bore in the world, namely, that at Rybuik, Upper Silesia, made by the German Government, for scientific purposes. The depth to date is 6,565 feet.

At Sperenberg, near Berlin, 4,173 feet were drilled for the purpose of obtaining Rock salt, and its great depth was reached without passing through the full stratum of Rock salt, which proved to be, so far as drilled, of the extraordinary thickness of 3,900 feet.

In Virginia, United States, a successful bore for water has recently been made to a depth of 5,060 feet, passing through, in the carboniferous rock, an enormous thickness of limestone.

The artesian wells in America (a description of which is given under the head of "American Bores") cannot be designated generally as deep bores, and few in England exceed 1,000 feet, and it is more particularly in London and its environs that success in obtaining large supplies of artesian water has been attained.

The United States Government have, however, given *carte blanche* to their engineers to put down a bore also for scientific purposes and regardless of expense. This will be commenced with a sufficiently large diameter to admit of very deep drilling, and will doubtless be continued as long as it is practicable to do so; the practical question being how far the anticipated heat of the earth, as the drilling proceeds to very great depths, will affect detrimentally the condition of the drilling tools.

As given under the head of "Queensland Bores" the Malvern Hills bore (private) is 3,948 feet deep, and is still in progress, and the Winton bore (Government) is 3,995 feet, giving 1,100,000 gallons per diem.

SECTION VI.

AMERICAN BORES.

AMERICA (the United States), with its wealth of science, mechanical skill, and unceasing energy and enterprise, was quick to adopt and improve the European system of artesian well-making, which in conjunction, where admissible, with the shallower—or sub-artesian—one of well-making, is common to the whole country. The discovery of great deposits of petroleum in Pennsylvania gave increased impetus to well-making, thousands of deep wells having been drilled to obtain the oil, and to such a perfect system has the business been reduced that the cost, duration, and success of the work can now be calculated with the greatest precision. As it is mainly due to the development of this petroleum industry in Pennsylvania that we owe the present practically perfect system of drilling for artesian water, and as petroleum and the coal oil-bearing rocks have at least indicated themselves in Australia, and as the oil has been recently found in New Zealand and may be waiting a large future general development here, a description of the mode of obtaining the oil in America (the mechanical means adopted for obtaining, which are identical with those used in drilling for artesian water) cannot, I think, fail to prove of interest to many readers.* It may be here remarked that in the subsequent discovery of large petroleum deposits in Canada and in other parts of the United States (notably Ohio) the Pennsylvanian system was adopted there.

"The records of mining, even for gold, present no parallel to the rapid development of such vast resources as those characterised by the American petroleum. Some forty years ago the first artesian oil well was drilled in Pennsylvania. From this small beginning has sprung into existence a mining business second only in extent and value to coal and iron, and to-day it constitutes the third estate among the chief mineral products of America. The presence of petroleum, or 'rock-oil,' manifested itself to the earliest settlers. To the native Indians it had been known and used as a medicine. Beyond this no use was discovered for the substance until its later development, when, in 1850, it was first utilised as an illuminator. The crude oil was obtained from salt wells, which had been sunk in the locality,

---

* The description is from the first edition of this book. Brisbane, 1883.

and gave great trouble and annoyance to the miners. The supply was very limited, and was pumped out of the wells at the usual depth from which salt water was obtained. The discovery of an important use for the article as an illuminator caused an extensive demand, and it became requisite to obtain it in large quantities. The sinking of a well similar to those of the salt regions was the commencement of the present petroleum development. In that well oil was struck at 69 feet from the surface, and public attention was then concentrated to the particular locality of the enterprise. Hundreds of people from all parts of the country flocked into the oil region and commenced sinking wells. These first operators contented themselves with drilling to a moderate depth, when they obtained small "pumping wells." The number of these increased rapidly until the latter part of 1860. At this time some reflective operator expressed the opinion that, as the supply of oil seemed to come from great depths below the earth's surface, deeper wells would reach the reservoir or source of supply, and greater quantities would be obtained. This theory was soon put to a practical test. A well was drilled to a depth of between 400 feet and 500 feet, and the 'third sand rock,' where the greatest supply is found, was reached. The result of this experiment astonished the operator and the country, too. When the drill penetrated the fissure or recess in the sand rock where the oil had collected the heavy drilling tools were hurled out of the aperture they had made through the earth with terrific power far above the top of the derrick upon the surface. The drilling apparatus was followed by a stream of oil the force of which was so great as to prevent the operators for several days from inserting into the well the casing necessary for it to flow through, and to control it.

"The general appearance of the oil country when the industry had become firmly established was one of unexampled industry. There were clusters and continuous lines of tall pyramidal derricks, engine houses, and board shanties, presenting the appearance of a town of that length. For this whole distance all was bustle and activity; the large flowing wells were spouting forth their oily treasure like huge whales, with a noise similar to that made by the steam pipe of a steamer heading against a current. Tall derricks arose on every hand, and amid the smoke and din were thousands of men busily employed in the various attendant occupations.

"One of the first of the 'flowing' wells was 'Drake's' well, and this was one of the wonders of the period. When oil was struct the production of this famous well averaged between 3,000

C

and 4,000 barrels of oil per day. The sight was probably the most remarkable of a life time. From the mouth of the conduit pipe the oil and gas came forth with terrific power propelled by nature's hydraulic forces. The oil of a beautiful dark olive green colour was dashed into spray against the side of the huge receiving tank, forming a prism of colours rainbow in hue, resplendent in the sun's rays, and beautiful beyond description. Here indeed was the Scripture parallel of the 'rock pouring forth a fountain of oil' enriching the fortunate owners, and giving a stimulus to operators throughout the whole region. The depth of this well was only 480 feet.

"Amid the moving panorama of oil wells, derricks, gushing fountains of oil, and wealth, energy and industry, of human hopes and alternate successes sufficiently varied to suit all the vagaries of an ordinary imagination was budding forth the germ of the magnificent petroleum development of to-day.

"The machinery used by the first operators was of the most primitive character, and the progress made in sinking a well was necessarily slow, and the labour tedious. That used at the present day has been gradually evolved during the petroleum development, is on the artesian principle, and is extremely simple in its parts and powerful in its working."

In a recent report on the proper location of artesian wells in a portion of the Western States of America furnished to the Government by the Department of Agriculture, Washington, in August, 1890, there are a mass of detailed reports by State engineers, geologists, and well-boring experts of the development in their several States or Districts of artesian water supplies forming probably one of the most valuable additions to our knowledge of the subject ever published. The immense area assigned to this elaborate examination embraces over 658,000 square miles. The reports may be condensed into the most salient features as follows:—The vast extent of the Dakota artesian basin is testified to by the enquiries and deductions of Messrs. Nettleton, Hay, Culver, and Bailey. Even a slight acquaintance with the chief features of the physical geography of the Dakotas would testify strongly to the probable permanency of the artesian water supply, which unquestionably is fed by the drainage of the Rocky Mountains to the north west. This drainage flow penetrates below the superincumbent stratum to the body of friable rock known as the Dakota sandstone, which appears to underlie the whole region. It is evident that the drill has nowhere more than penetrated a few inches of this water-bearing and conserving stratum. The altitude, the general trend of the land, and the formation and character of the great hydrological or river area which intersects it, give weight to the deductions that are made as to the extent and permanency of this remarkable artesian basin. There are found within it about 150 high pressure artesian wells. There are also found in South Dakota several hundred *flowing* wells, whose supply is evidently from sources not identified with the greater artesian basin. In North-east Dakota, in the hydrological basin of the Red River, claimed by geologists to be the seat of an ancient lake, there can be found over a thousand small flowing wells, whose waters are used largely for farming and homestead purposes in garden and other small irrigations. *No diminution of pressure* is anywhere reported. The source of this supply is from the upper beds of glacial drift.

Controlling Geological Conditions.

Professor G. E. Culver, the State Geologist for South Dakota, has made a report of singular interest—cautious, but still comprehensive. It shows that the great Dakotan basin—a broad, low, synclinal with a north-east trend—has a width of 550 miles and length of 700 miles, and that the controlling factors are:—

1. The position of the pervious Dakota sandstone between two impervious beds.
2. The flexing of the whole series forming a low, broad synclinal.
3. The tilting of the beds, giving a long easterly slope, with a slight rise near the eastern border of the basin.
4. The exposure of the Dakota sandstone on the western rim, where mountain streams and drainage must cross it.
5. The overlap of the Colorado shales and clays on the eastern border, sealing in the waters and allowing an accumulation.

The Red River Basin, in the north-east portion of North Dakota, obtains its water from rocks much older than the Cretaceous. There are in this valley a great number of shallow wells of moderate flow and pressure, the water of which is found just beneath the glacial clays in recent beds of clay and gravel. A belt of wells somewhat similar to this has been found in the basin east of the James River, in South Dakota. They are shallow in depth, and of character common to the wells of Nebraska and Kansas. An important examination was also made of the geology of South Dakota lying west of the Missouri

River, and of that portion of Wyoming lying west of the Platte River and east of the foot-hills of the Rocky Mountains, by Professor G. E. Bailey. It is claimed in his report that the chief conditions requisite to artesian wells exist throughout the south-western portions of South Dakota. These conditions are :—

1. A porous stratum furnished by the Dakota sandstone.
2. Impervious beds above and below the Dakota sandstone.
3. A high fountain head in the Black Hills and mountain ranges along the western edge of the State. In order to obtain an artesian basin, geologically speaking, the following conditions are necessary, all of which are found in the Dakotas :—

(1) A pervious stratum to permit the entrance and passage of the water.
(2) A water-tight bed below to prevent the passage of the water downward.
(3) A like impervious bed above to prevent the escape upwards.

The first is furnished by the Dakota sandstone, a bed of from 250 feet to 400 feet in thickness ;* the second by the Jura-Triassic system of rocks, which are immediately below the Dakota sandstone and directly above the Carboniferous limestones. The rocks above furnish what is known to geologists as the Colorado group. Other conditions needed to complete a basin are all found in the trend or inclination of the strata, a proper exposure of the porous stratum for collecting the water, and an adequate rainfall with absence of escape from local cause. These physical features are all found in the Black Hills region, and, as Professor Bailey declares, are accompanied with a gentle inclination of the strata to the east, thus making the entire area from foothills of the Black Hills region to the Missouri River an artesian basin.

Professor Bailey, after a geological examination of the State of Wyoming, assumes that in the eastern portion of the State a number of small artesian basins may be found, wherein good water could be obtained from the Dakota sandstone, which stratum bears the same character as it does in Dakota. The Tertiary rocks and sandstones are open and porous, and below them is a good water-tight stratum, while above the stratum is impervious. These Tertiary rocks are generous water-bearers, and though the pressure of their supply would be very low, yet the water they contain can be easily reached, and by mechanical power brought with economy to the surface. The geologists

---

* Note the great thickness of the Dakota sandstone (the equivalent of the Queensland braystone of Mr. Jack). This shows what may exist in Australia.

regard "the development of the water in the Tertiary strata as of the utmost importance," and urge a more thorough investigation. Professor Hicks, of Nebraska, offers an intelligent presentation of the stratigraphical conditions which determine in his field the direction, the flow, and the pressure of subterranean waters. Nearly all such waters in Nebraska " have a tendency to rise in wells and sometimes throw out at the surface." He considers the rainfall the ample source of this supply. The aggregate volume of such precipitation is enormous. It amounts, he says, "to more than 100,000 cubic feet in each acre, or 5,000,000,000,000 of cubic feet annually for the whole State." The mean discharge of all the rivers of Nebraska is estimated at nearly one-fourth of this great total. Evaporation and seepage absorb the balance. However, the "underflow along the incline bodies of porous rocks is undoubtedly a more important source of moisture than all the rivers which enter the State." The existence of great levels of this sheet water has been determined and demonstrated. Many streams have a steadier volume in their channels than the precipitation and their incline warrant. It is believed that they are eroded deep enough to receive a supply of this undersheet flow. The water-bearing strata of the State are geologically known as the Perma-Carboniferous. The youngest rocks of the stratum are the most valuable. Artesian flow has been obtained from this stratum, and, contrary to usual experience, the limestones appear to yield more than the sandstones of the same area. The land and sandstones of Cretaceous origin, which lie upon the Carboniferous formation, are also found to be good water conservers and yielders. A very much larger area can be brought under cultivation by the use of wells, from which the water is pumped. A notable fact in regard to the Nebraska undersheet water is that in the western part of the State it moves under considerable pressure and when opened rises rapidly to or very near the surface.

Field Agent Gregory, in closing his careful report, reviews conditions of the Great Plains region from the northern boundary of Nebraska to the southern boundary of the Indian Territory. He found within his division about 200 flowing wells, with several hundred more in which the water rises, but does not reach the surface. Accompanying his report are the records of numerous springs unquestionably artesian in character; but the great source of supply in the central region must be looked for in the undersheet water found in the gravelly strata below the alluvial surface.

In Colorado there are four distinct artesian basins reported; that at Denver, the capital, is at present the most thoroughly developed. Water is found from 65 feet to 1000 feet deep. A decided success has recently been made in serving market gardens of from eight to fifteen acres each. There are about 350 wells in this basin. The Greely basin is one of deep wells; the average depth is from 1100 feet to 1300 feet. The Pueblo basin is a similar one. West of Pueblo, at Florence and Cañon City, along the drainage basin of the Upper Arkansas, several heavy flows of water have been struck. The most remarkable basin found in Colorado is situated just outside and west of the line of this investigation. Within three years, and largely during the past year, over 2000 flowing wells have been sunk in the San Luis Valley or basin, which contains 16,000 square miles. The water is used most extensively for domestic and stock purposes, but a considerable number of the farmers are now entirely dependent for their irrigation supply on these wells. In the village of Monte Vista there are some 90 wells. The Empire Farm at Alamosa is supplied by 40 wells. It is probable that from 10,000 to 12,000 acres of land were wholly or partially irrigated by these wells. No diminution of flow has been perceived.

Field Agent F. E. Roesler makes a valuable report from Western Texas. He states there are 700 flowing wells in his section. Five flowing wells at Waco flow jointly 5,000,000 gallons per diem, and a notable feature of them is that the water-bearing stratum is found at a depth of at least 1200 feet below the level of the Gulf of Mexico. Another remarkable feature of this section of Texas is the supply of water from springs, among them being some of the largest in the world. At Lampasas, from one spring, a flow of 2,400,000 gallons per diem is obtained. Mr. Roesler's view of the wells developed and the use of the water thereof in horticultural pursuits demands careful attention. The speculative references he makes as to the existence of undersheet or flowing water below the surface deserve the consideration that belongs to the views of a careful observer of large experience, knowledge, and good judgment to sustain his deductions.

In Utah about 2000 bored wells have been sunk. Evidence goes to show that no diminution of pressure has yet occurred. In Nevada similar valuable work has been accomplished. California, however, most prominently illustrates the extent of artesian water and its economic value. There are upwards of 3000 artesian wells in this State irrigating up to 1000 acres each. They exist from San Diego to Shasta County, and while their flow is not one-tenth utilised probably 60,000 acres are at present served. Their possible development in arid regions is almost illimitable.

Professor C. W. Hall, of the University of Minnesota, writes as follows on "The Geological Conditions which Control Artesian Well-boring in Minnesota":—*" The sandstones and limestones which are exposed along the river gorges can be followed from one gorge to another, or from one deep well to another, over nearly all South-eastern Minnesota, and are known to occur in Wisconsin and Iowa. The sandstones are the water-bearing strata, and this persistence is important to the well-borer. It enables him to calculate to a very close figure the depth in any part of this area to which he must bore in order to reach a formation that everywhere, according to experience, yields an abundant supply of water." After defining the principles of artesian water, Professor Hall continues :—" There must be a sufficient freedom from fissures, faults, and dikes to ensure a steady flow without great loss of water from the rainfall district to the region of the wells. It is not necessary that the porous stratum be a sandstone further than the natural qualities of the rocks themselves determine the question. No other rock species is sufficiently porous to permit a free flow through it of large quantities of water, save by fissures, and these form an unreliable passage way, even more likely to cause defeat than to ensure success in the search for water. The sandstones of Minnesota can very easily be parallelized by means of the deep wells already bored, and their general relations to each other and to the interbedded limestones can be made out."

As I have stated, the petroleum industry gave a great impetus to boring generally and for subterranean water especially, the machinery and system evolved from boring for oil being in every way fitted for obtaining artesian water. The importance of the results obtained being fully recognised, led to a careful collection of information by the Government Departments, and a surprisingly extensive fund of knowledge is now available. A notable feature of these reports is in their showing that valuable sub-artesian water may be found and utilised in the deeper water-bearing country, the geologists regarding "the development of the water in the Tertiary strata as of the utmost importance," and urging more thorough investigation. This is in a region of high development of the deeper artesian water, and further shows in

---

* Bulletin Minnesota Academy of Sciences. Vol. 3.

how thoroughly efficient and practical a manner every available source of water supply is made use of in America. Artesian wells in that country are reckoned by thousands, extending from Montana and North Dakota to the southern portions of Texas. Besides in the States of North and South Dakota, Wyoming, Nebraska, Colorado, Kansas, Idaho, California, Nevada, New Mexico, and Texas, in which their numbers are almost incredible; in the Northern and Eastern States, likewise, an immense number of wells have been bored. As will be seen in the Section, "Permanence of Supplies," the total number for the whole country, as given by official statistics recently published, is nearly 17,000. These wells at an average depth of 1000 feet per bored well, would give about 3240 miles of boring. Surely such a result as this, achieved by individual efforts in the development and utilisation of underground water supplies in so highly experienced and practical a community as that of the United States, speaks volumes as to the economic value of that water, both deep and shallow, and points out the possibilities in the same direction in Australia.

AMERICAN BORES.

Section showing the dip of the Artesian Water-bearing Rock—the "Dakota Sandstone"—towards the Sea Level.

SECTIONAL VIEW, showing relation of water bearing rock to the surface.

Horizontal Scale, ¼ in ·10 Miles
Vertical „ ¼ „ ·100 Feet

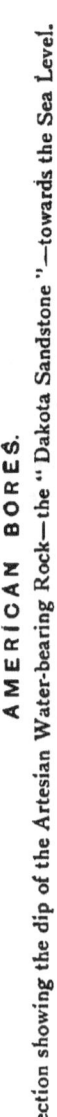

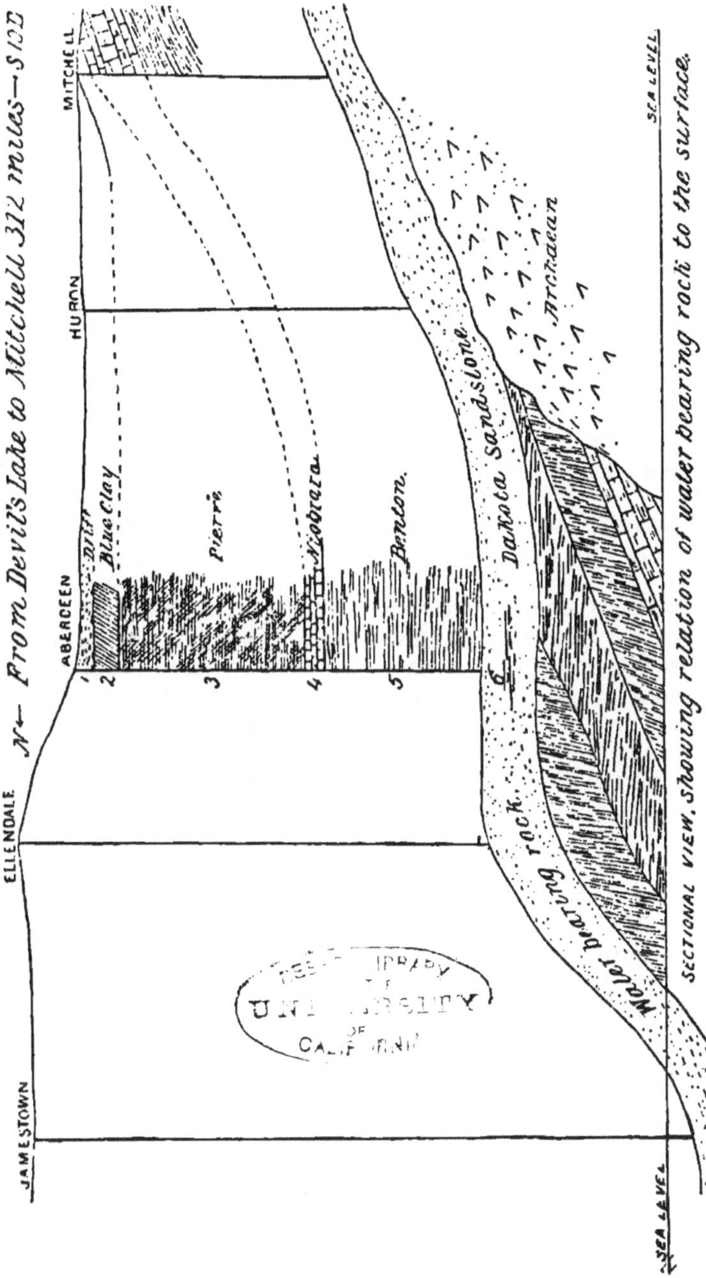

AUSTRALIA.

SECTION VII.

VICTORIA.

THE Victorians have, with their usual well known enterprise, been assiduous in testing their colony for artesian supplies with more or less success. One of the first artesian wells made in Australia was that put down at Sale, in Gippsland (private), in 1880.* The depth of the well was 234 feet only. It was commenced with 4 inch casing for 100 feet and reduced to 2½ inch casing to the bottom. The flow rose 16 feet above the surface, and the supply was 36,000 gallons per diem. The water was somewhat saline; but a fact worth recording is, that the water after exposure and its attendant aeration became fresh.

Boring has been largely, and is still being, carried on in Victoria in search of water in the Post-tertiary and Tertiary deposits which occupy about one-half of the area of the colony. A series of borings has been made in the Mallee country, but hitherto with no very marked success. Upon a line of bores laid down running from Nhill to the river Murray, a series of bores have been made, the deepest, that at Netherly, being 2200 feet. This was made a considerable distance into the bed rock, and then abandoned. Similar operations have been carried out to the eastward from Donald to the river Murray, but without the desired result. Artesian water has been struck at Mordialloc with a limited flow of fresh water. By means of pumping the outflow amounts to 65,000 gallons per diem. Fresh water has been struck in many of the bores, but the supplies are much less than those tapped by their northern neighbours. The character of the strata passed through are, as before stated, those pertaining to rocks of the Post-tertiary and Tertiary ages, as no equivalents to the extensive Cretaceous water-bearing formations of the northern colonies or South Australia have, as far as I can learn, been so far met with.

The Tertiary rocks of the north-western district, where boring is being carried on, rest on rocks of a Primary age, principally Silurian, which constitute, except in places where the subjacent granites and trappean rocks have been laid bare, the bed rock

---

* Two years after my first communication to the Press—the Melbourne *Leader* upon " Artesian Wells for Australia."

of the colony. A private bore has been put down in Melbourne into the bed rock; but the departmental practice is to cease boring when rocks bearing evidence of a Primary age are reached. Boring for water appears to be more successful in the coastal districts—the reverse of what occurs in Queensland.

The Victorian Government have expended since 1886 nearly £60,000 in the search for subterranean water, proving the estimated great value placed upon the possession of it, and the admirable enterprise, although not as yet so entirely successful as Australians could wish, of the Victorians in this direction.

SECTION VIII.

SOUTH AUSTRALIA.

SOUTH AUSTRALIA must be given a foremost place in the extent of its artesian areas. A geological examination by the Government Geologist, Mr. Henry Y. L. Brown, shows that a very large area of the colony may be said to embrace a wide Cretaceous basin, extending from the Queensland and New South Wales borders to the outcrop of bed rock near Farina, the limits on the north and west being as yet undefined; but so far as I can ascertain it embraces an area of nearly 100,000 square miles.

The importance of wells and artesian bores has long been recognised; and I have a very satisfactory recollection of South Australia being to the front in the search for subterranean water in the earliest times of its initiation in Australia. Sir Thomas Elder was one of the first far-seeing enterprising colonists who took up the search. He fitted out a well-boring expedition about the year 1881, using camels to carry the boring-plant to his back stations.

Various efforts have for many years been put forward by the Water Conservation Department, to provide a supply of water on stock routes and at various centres, and most valuable work has been done; and the records kept by the Department in the collection and classification of strata passed through in the various bores; lithographed sections of the bores and other carefully compiled information, evince remarkable care and completeness, which it would be well if it were followed generally in similar undertakings. The borings themselves have been carried on in a very systematic manner through very repellent strata, such as hard limestone and well into bed rock—granite formation — with persistent and most commendable enterprise on the part of those in charge of the department meriting every success.

Owing to the enormous extent of territory the borings are much scattered.* The bore at Wilmington—the earliest of the bores attempted, resulted in a sub-artesian supply being tapped at a depth of 350 feet. On the two following pages is a list of bores completed and in progress.

---

* *Vide* map published by the Department, a copy of which, with particulars of borings, has been very courteously sent to me by the Chief Engineer, Mr. A. B. Moncrieff.

South Australian Bores.

List of Bores put down by the South Australian Government in search of Water.

Locality.	Depth. feet.	Galls. per diem.	Results.
Hund. of Coglin, Travellers' Rest Bore	908	1,200	Good water.
Hund. of Schomburgh	365	25,900	Stock.
„ Dublin, Windsor Bore	391	...	Stock, large supply.
„ Boolcunda	473	...	Stock.
„ Cameron, Percyton Bore	930	11,000	Good stock.
Hergott No. 1 Bore	353	6,000	Good water, artesian
Mirrabuckina	1635	...	Stock.
Hund. of Morgan	575	28,000	Good.
„ Dublin, town of Dublin	212	20,000	Slightly brackish
„ Coonato, Hammond	288	18,000	„
„ Mudla, Wirra, Wasley's	520	...	Stock.
Hergott No. 2 Bore	342	2,700	Good, artesian
Coward	308	1,200,000	Stock, artesian
Strangways	365	1,200,000	„
Nullarbor Plains, Roberts' Well Bore	777	68,000	Good stock
Coonalpyn, S.E.	840	...	Stock.
Hund. of Inkerman	524	...	„
„ Willochra, Bruce No. 1 Bore	233	10,000	Good artesian
Dulkannina Bore	855	...	Stock
William Creek No. 1 Bore	272	52,800	Good stock, artesian.
Hund. of Willochra, Bruce No. 2 Bore	385	3,600	Good artesian.
Nullarbor Plains No. 2 Bore	815	19,200	Stock.
Tintinara, S.E.	253	4,300	Good artesian.
Ke Ke, S.E.	666	...	Stock.
Anna Creek	988	...	„
Hund. of Maitland	266	...	„
Piarooka	914	...	„
Hund. of Dublin	583	...	„
„ Palmer, Kingswood Bore	330	28,800	„
Bedabore, N.W., Port Augusta	1099	...	„
Albala Karoo, near Eucla	1084	...	„
Emu Flat, S.E.	268	31,296	Good.
Maitland Township	232	...	Stock
Hund. of Tiparra, Weetulta	112	...	„
„ Cunningham	502	...	„
Lake Mulyungarie No. 1 Bore	1030	...	„

Artesian Wells:

Locality.	Depth. feet.	Galls. per diem.	Results.
William Creek No. 2 Bore	228	3,840	Slightly brackish, artesian
Hund. of Boolcunda	745	19,200	Good stock, artesian
Lake Harry	1360	100,000	Good stock, artesian.
Richman's Valley	1202	28,800	Slightly brackish
Gumewarra, West Coast	1277	...	Stock.
Lake Mulyungarie No. 2 Bore	1909	...	,,
Lake Frome	1412	...	,,
Frome Downs	691	...	,,
Croydon	2296	...	Fresh water
Nullarbor Plains No. 5 Bore	669	288,000	Good stock
Oodnadatta, O.T. Line	1571	270,000	Good artesian
Snowtown	299	...	Good.
Millendilla	231	...	,,
Gawler No. 1	126	...	,,
,, No. 2	140	...	,,
Boolcounda	745	...	,,
Hammond	238	...	Stock
Eurelia	242		Good stock
Stephenson	230	...	,,
Bruce No. 1 Bore	233	...	Good
,, No. 2 Bore	385		,,
Nankara	250		Stock
Wilmington	646		Good
Stirling North	462		,,
Farina	322		Stock
Hamilton Creek	...		In progress.
Blood's Creek	...		,,
Charlotte Waters (N.T.)	...		,,
Corrie Appa	570		,,
Lake Crossing	...	...	,,
Koperamanna	...	...	,,
Alexandria (N.T.)	1664	...	,,
*Two Wells, 25 miles N.W. from Adelaide	...	...	Very large flow of good water.

Some of the most successful bores are those named at Hergot, Coward, Strangways, and Mungamurtee, in which water was obtained at the respective depths of 342, 308, 365, and 237 feet, with outflows of 1,000,000, 1,200,000, 1,200,000, and 52,000

---

* This is the first satisfactory bore that has been put down within 100 miles of Adelaide, and speaks well for future operations in the district.

gallons per diem in each case. These large outflows at such comparatively shallow depths are extremely satisfactory, and in their similarity to the shallow artesian borings at Uanda Station (Queensland), and in New South Wales and the United States prove that artesian water does not invariably lie at great depths, but that in some localities it is easily obtainable.

In taking a broad view of artesian boring in Australia, seeing the extent of the development, rapidly increasing, of artesian supplies in the adjoining colonies and that the geological formation gives a predominance to the water-bearing rocks and areas in South Australia, the strange feature of the experience, so far as this province is concerned, is that, having got so far as to demonstrate the peculiar applicability of the artesian system to the northern country, they have gone so little further. Low prices of wool and uncertainty of tenure have doubtless had a good deal to do with the apathy shown by private people ; but it seems apparent that the Government would do well to consider whether, as a matter of policy, special inducements should not be offered for the encouragement of those in places far distant from other proved areas of artesian supply by making borings at the Government expense. At the same time it is very satisfactory to find that many bores have recently been completed or are in progress in various waterless parts of the outside country, as will be seen by the list given.

As in the other colonies, the practice has been to use mostly the shallow boring machines and the diamond drill. The practice latterly has been, however, to use the sole mode adopted in Queensland and New South Wales for artesian boring, viz , the Canadian pole or cable system.

Section IX.

WESTERN AUSTRALIA.

MR. ERNEST FAVENC, in his recently-published "Waste Lands of Australia," writes:—" At the present time more than one-quarter of the Australian Continent may be said to be unproductive, and the question of its probable future is one naturally pregnant with importance. The utilisation of about 800,000 square miles of additional territory must mean a considerable augmentation of population and revenue. It will, of course, at once occur to the reader that these waste lands owe their present neglected condition to their forbidding character: the absence of surface water, the irregularity of the rainfall, and the distance and difficulty of carriage and transit. On the other hand, we can look forward to the success that is attending artesian boring, the steady revolution that has taken place and is still taking place in our estimate of the capabilities of inland country, and the constant advance and extension of railway communication as tending to counterbalance the natural disadvantages. Thus our position with regard to dealing with these hitherto despised tracts is, in consequence of these united forces, increasing every year.

"For all practical purposes we may consider the whole of Australia as explored. The still unknown patches are isolated, and their value can be readily estimated by the physical character of the adjoining country. It may, therefore, be accepted that we are fairly acquainted with the nature of the country which in future we shall have to utilise, and are prepared to deal with from, to a certain extent, the standpoint of experience.

"Amongst these waste lands we may count upon finding many large areas of available country, probably waterless, which will become, by means of the drill and conservation, outposts by means of which the remaining portion will be gradually reclaimed.

"The first invasion of the existing interior desert will, as usual, take the form of pastoral occupation of the large tracts of dry glassy plains known to exist from having at various times been crossed by the many parties that have now traversed the no-man's land that intervenes between the border Stations of South Australia and those of Western Australia. This occupation, dependent on the subterranean water supply, augmented by the somewhat slender resources of conservation, will wait on the further development of railway enterprise. Our present improved mode of station management will also give the country a start under favourable auspices.

"Although, with regard to surface water, the unsettled tract which usurps the greater half of Western Australia is admittedly deficient, enough evidence is forthcoming to prove the existence of a considerable underground supply at no great depth. Springs are to be found throughout the interior, some, it is true, not remarkably palatable to man, but most of them passable stock water. So far as pastoral occupation has pushed to the eastward from the Western Australian coast, water has been obtainable in a limestone formation at a depth of from ten to twelve feet, and the supply is constant and reliable. There is no reason to suppose that this feature, which has been traced on to the inland water-shed, should not continue right across. This would provide the necessary requirements for pioneer stock, and serve the purpose of breaking in the land to settlement. Artesian boring would follow in due time."

An interesting account of water existing in the limestone formation in Western Australia is supplied by Mr. Gillett in a description of the country passed through in his exploration in Western Australia between Northam and Eucla in 1887, and published in the "Proceedings of the Royal Geographical Society of Australasia." Vol. X. No. 3. Mr. Gillett describes the Hampton Plains as covered in many places with flowers, and in other places with a rich variety of grasses, and then goes on to say :—" After entering upon the limestone formation the country, which had hitherto consisted largely of open forest, opened out into grassy and salt bush plains, which extended to Eucla, and should be well adapted for pastoral purposes. There we discovered many caves or holes, which appeared to penetrate to a great depth into the earth, and which I have reason to believe are to be found in other parts of the limestone country. We examined two of these caves. On approaching the edge of these natural wells we were astonished to hear, proceeding from a vast depth, a noise as it were of a mighty torrent, a dull sullen continuous roar ; pieces of rock thrown into the abyss returned no sound beyond the rebounding from side to side until ultimately lost to hearing, but no splash could be heard; yet still the ever varying roar of water continued, tantalising enough to those who, whilst they heard the sound of rushing waters, had still no means of obtaining it." Somewhat corroborative of such subterranean rivers is a statement credited to Captain Von

D

Schmidt, of the schooner "Grace Darling," trading between Port Adelaide and Eucla, that in the neighbourhood of Streaky Bay he saw at low tide an enormous volume of fresh water rushing from beneath the cliff and preserving its freshness for some distance out to sea. He further stated that he mentioned this to a few settlers in the district, who were too apathetic to take any interest in the matter.

This underground flow and escape of fresh water into the sea elsewhere, and that often in large quantities, is evidenced by Humboldt's description of a submarine spring in the Gulf of Mexico, some considerable distance from the South American coast, which converts so large a space into a fresh water lake that it is inhabited by the fresh water Cretacea found in the Orinoco; also an abundant spring of fresh water in the Indian Ocean 125 miles from Chittagong and 100 miles from the coast of the Sunderbunds. ("Humboldt's Travels.")

It is evident that this subterranean water cannot be deeper below the surface of the plains than the altitude of the plains above sea level. The plains are seldom more than 800 feet above the sea so that as the subterranean river must have a fall towards the ocean, it would probably not be more than about 500 feet below the surface; perhaps not more than a few hundred feet. The phenomenon of underground rivers is by no means uncommon in limestone country, and the volume of water conserved and travelling to its lowest level, the ocean, is in this formation especially no doubt enormous.

In the Report of the Department of Mines, Perth, 30th June, 1894, is the following which shows that fresh water may be obtained in large quantities even in the auriferous (mining) districts:—

"Country between Broomehill and the Dundas Hills and the Mines in that neighbourhood.

"The water question of late years has caused great trouble, and the rainfall has been lighter than usual, and owing to its elevated position wells have, as a rule, to be sunk to a considerable depth into the solid rock to assure a permanent supply of water. Many tanks have been excavated, but as the gullies have not run the last few years these have not been filled. In the bed of the Pallinup River there are several nice pools. There are large numbers of fresh water lakes and swamps.

"On the road between Magitup and Terramungup is a celebrated natural curiosity, called the "Night Well," in reality a fissure in the granite rock in the bed of the creek, which during the summer months, is dry during the day, and full of water (often to overflowing) during the night; whilst during the winter it is always full, but is mostly covered by the water of the gully. The intermittent nature of this spring is evidently due to the changes in the temperature, for after very hot days it is either not filled at all, or not until nearly morning, whilst as a rule, the water rises at from 10 to 12 p.m., disappearing again suddenly in the morning, as soon as the day begins to get hot. The immediate cause of the water rising is due either to the expansion and contraction of the rock with the rise and fall of the temperature, which would cause the fissure to open or close, or upon the other hand as the water is probably derived from the drainage of the high sand plain to the northward, the water being sucked up by the heated upper sand beds and plant roots, particularly if these beds are not of any great thickness above the water-bearing beds.

"Notes on Water Supply.

"It is not easy to treat this subject generally. Each locality, where the number, size, and richness of the gold deposits will enable the place to become a permanent mining centre, has its peculiar features, which in every case has to be specially considered.

"The annual rainfall, especially in the southern and south-eastern portions of the gold fields is apparently not a small one, but at the same time large accumulations of waters on the surface are of rare occurrence. A climate with tropical torrents, or with an annual melting of snow masses at spring time, would soon completely alter the features of these regions; large masses of water would be able to cut channels, and so effect a drainage of the country towards the sea; the soil would become leached of the salt, and in consequence of the removal, separation and redisposal of the detritus by water, the surface would become less pervious for the rain water; creeks and rivers would enliven the country, and flora and fauna would develop. As man has no power to alter the climate, he must seek out and employ obtainable means for the purpose of reaching the desired effect.

"Up to the present time successful endeavours to obtain the necessary water supply have followed two distinct systems. One is the collection and preservation of surface water in tanks, and the other consists in the sinking of wells into the strata of elevated areas of the archæn formation. These strata are the unremoved and accumulated detritus derived from the decomposition of the archæn genesis—granite rocks.

"*Both systems are capable of a great development.* For collecting water on the surface only certain places are favourable. The principal conditions for a catchment area are—basin form, a clayish surface soil, and an elevated situation above the level of salt lakes and pools. In almost every case it would be necessary to prepare the surface of the catchment area, or parts of it, artificially, so as to enable it to retain and conduct the rain water. Reservoirs of solid construction would have to be erected, and covered so as to reduce evaporation to a minimum, large and elevated areas consisting of unremoved and accumulated detritus, and derived from the underlying archæn rocks, border the auriferous belts of country towards the east and west. In those areas considerable supplies of fresh water could be secured. There have been valuable wells sunk in the strata of accumulative decomposition which usually surround rock outcrops. Here it is not necessary to retain the rain water on the surface. The stratum of solid rock underlying the detritus forms the collecting and conducting plane. The surface water enters, and percolates the strata of accumulative decomposition, but in their downward tendency they have to follow the way designated by the flat inclined underlying solid rock. The configuration of such rocks contains depressions and elevations. If basin-formed and closed depressions occur they offer subterranean reservoirs. To procure water supplies from such areas systematic proceedings would be the first condition of success. Boring tests and geological surveys would have to form the basis for the adoption of the plans of operation."

The surveys are now in progress, and fresh water has already been obtained by boring notably on the road from Southern Cross to Coolgardie; at Hampton Plains, and various places in the mining districts and also along the base of the Darling Range where—bearing out the opinion expressed some years ago by the Government Geologist—a bore recently made to a depth of only 111 feet gives 8,900 gallons per diem of fresh water. The Government intends to open a stock route between Murchison and Coolgardie, and will also put down bores for artesian water around Coolgardie. They are to be made to a depth of 3,000 feet unless water is struck.

It is to be hoped that the Western Australian Government will, as soon as the Geological area—undoubtedly a large one—of the water-bearing rocks is established by the Department, prosecute on an equally liberal scale the utilization of the subterranean water supplies as it will the carrying on of the necessary railways and other public works, and in this new departure in the expenditure upon public works evince a sounder policy than the older colonies of Australia have done in the past. As a shrewd pioneer pastoralist remarked to me years ago:—" Railways are all very well, but if a portion of the money had been spent in putting water upon the lands more good would have been done. A man's property," said he," might be 'cobwebbed' with railways, but without a water supply it would avail him little !"

Section X.

NEW SOUTH WALES.

As was to be expected of the parent colony, with its advancement and wealth, and large areas of water-bearing rocks, which its geologists have shown that it possesses, much valuable work has been done, and is in progress both by the Government and by lessees in developing its wealth of artesian water.

In the last Report (1891) by Mr. J. W. Boultbee, of the Water Conservation Department, Mr. Boultbee says:—" The question has been for years before the Department, attention being in the first instance drawn to it by a successful artesian well sunk on Kallara station by David Brown, Esq., in 1879. This well was sunk in proximity to a mud spring, and at a depth of 140 feet artesian water was tapped that rose 26 feet above the curb, and has been flowing without intermission ever since. The first essay made by the Department for artesian water was in 1884, although artesian water had been previously tapped in a bore for coal near Lake Macquarie, and under the direction of a former Superintendent of Drills (Mr. Henderson) a bore was put down at Goonery, an out-station of Toorale run, on the Bourke-Waanaaring road, 51 miles from Bourke, to a depth of 89 feet, at which depth water was struck, which rose 8 feet above the curb at the rate of 1,000 gallons per hour. Other bores sunk in the same locality proved successful so far as reaching artesian water was concerned, but so impregnated with saline matter as to be useless. In 1885 and 1886 the present Superintendent of Drills (Mr. Slee) sunk a bore at the 75-mile peg, now the Tinchelooka bore, upon the same road, to a depth of 960 feet in the face of great difficulties due to the drought, &c., the water from which rose to a height of 20 feet above the curb, at the rate of 33,000 gallons per diem, the quality not being first-class.

"In 1885 in a bore for coal at Ballimore near Dubbo, artesian water of a mineral character was struck at a depth of 550 feet, rising at the rate of 1,000 gallons per hour to a height of 30 feet above the curb.

"In 1887 at 101½ miles from Bourke, a bore (the Cuttaburra bore) was sunk to a depth of 965 feet, at which depth water was tapped and rose to a height of 8 feet above the curb, yielding 22,464 gallons per diem, the water being of an inferior quality, due, it is supposed, to the salt water struck near the surface not being sufficiently shut off. These bores were all sunk in pursuance

LILA SPRINGS STATION, WARREGO DISTRICT, NEW SOUTH WALES.
Depth, 1,729 feet; Flow, 4,000,000 gallons daily; Temp., 124 deg. Fah.

of a recommendation made to the Government in 1880, after the successful effort made by Mr. David Brown at Kallara, by Mr. Wilkinson, the late Government Geologist, in conjunction with the Inspector of Public Watering Places, Mr. Gilliat, and the Chief Inspector of Stock, Mr. Bruce, to put down a series of bores for artesian water with a view of opening up a new road and well-watered stock route from the Mount Brown Goldfield to Bourke to lead the northern traffic to the railway at Bourke. During the progress of this work Mr. Geological-Surveyor Brown was engaged in examining the extensive water-bearing country in the Albert and Warrego districts ; and this work has been continued by Mr. Geological-Surveyor Anderson, who has so far been able to determine very largely the probable southern limits of the Cretaceous or artesian water-bearing formations."* The inadequate appliances, viz. :—The Tiffin, and Wright and Edwards' augers, at the command of the Superintendent of Drills, prevented rapid progress in the work. The success in Queensland drew attention to the contract system of carrying out this work, and acting upon the advice of Mr. Slee, the then Secretary for Mines, Mr. Abigail, issued instructions for the calling of tenders for a bulk amount of 30,000 feet of boring.

" With reference (further says Mr. Boultbee) to the origin of the artesian water in New South Wales, the geological map of the country shows that the area within which the artesian water lies, is shut in the south-east by the great impervious barrier of Palæozoic rocks which constitute the main dividing range ; but to the north and west communicates with and forms part of the artesian water formations of Queensland and South Australia, possibly on the south-west the artesian basin is hemmed in by a low ridge of impervious Palæozoic rocks, extending from near Dubbo by way of Nymagee and Cobar to the Barrier Ranges. Possibly, however, a narrow inlet may exist in this sunken ridge, approximately underlying the present channel of the Darling River and allowing the water in the artesian beds to escape underneath the Tertiary deposits of the Riverina district to the ocean at the Coorong coast near the mouth of the Murray River. Traced northwards into Queensland the artesian basin is still bounded on the east by the same main dividing range and its various offshoots as far north as the Gulf of Carpentaria. . . . The artesian basin is obviously completely closed along its eastern margin, but may discharge its waters into the ocean either in a northerly direction at the Gulf of Carpentaria or westerly ; and

---

* See more recent Report (December, 1894) of the present Geological Surveyor, Mr. E. F. Pittman, page 40.

then south-westerly, following approximately the course of the river Darling into the ocean near the Coorong coast. There is a strong probability that the artesian waters of the Cretaceous basin, are subject to a slow but constant underflow, which enables the water to circulate, and eventually discharge its saline matter into the ocean, for it is on this property of circulation that the freshness of it chiefly depends.

"It is a well-known fact that rain-water which has percolated into the Tertiary formations, which overlie large areas of the Cretaceous formation of our western plains, rapidly becomes salt in those areas where it remains stagnant, but in localities where water-bearing beds of the same formation overlie the porous beds of the Cretaceous formation, the water found in the former is fresh owing to its being able to circulate. The source of the water in our portion of the artesian basin is obviously the rainfall of the upper portion of the Darling catchment, and particularly that which drains into those portions of the Dumaresq, Gwyder, Namoi, Castlereagh, Bogan, and Macquarie Rivers, which overlie impervious bed-rock."

The most recent report (December, 1894) of the artesian water-bearing areas of New South Wales is that of the present Government Geologist (Mr. E. F. Pittman), who made an examination of the north-western portion of the colony with the special object of ascertaining the areas occupied by the Cretaceous or artesian water-bearing rocks. The area examined lies to the west of the Paroo River, and is bounded on the north by the Queensland border and on the south by a line joining Broken Hill and Wilcannia.

After alluding to examinations made by former Government geological surveyors, Mr. Pittman says :—"Briefly summarised, their reports are to the effect that this territory consists of several areas of Palæozoic rocks (intruded by dykes of granite, diorite, &c.), which contain deposits of such metals as gold, silver, copper, and tin, and which are flanked or surrounded by Cretaceous or water-bearing sediments covered in places by drifts and sands of Pleistocene and recent origin. The examination of the district recently made by me convinces me that the Palæozoic areas shown on our geological map must be considerably reduced, and that, on the other hand, the area occupied by the Cretaceous or water-bearing rocks is much larger than was previously indicated.

"Perhaps the most important conclusion at which I have arrived is that the artesian basin has probably a much further extension southwards than had been previously assigned to it.

It has hitherto been considered that the southern boundary of the Cretaceous basin was formed by a bar or buried range of Palæozoic rock stretching westward from Cobar through Wilcannia to Scrope's Range. At Wilcannia the rocks forming the supposed bar were regarded as Devonian, and this opinion appears to have been formed on lithological evidence only, as there is no record of any Devonian fossils having been found *in situ*, nor of any geological section in which the relation of these Wilcannia sandstones is described with regard to older sediments. But in my opinion the lithological character of these sandstones points to their being of Mesozoic rather than of Palæozoic age, and the small amount of geological evidence that can be obtained from a surface examination seems to strengthen that view. Deposits of hard sediments which I observed at certain localities such as the west of the Koka Range, at Kooningsbery, at the western side of Woychugga Lake, at the western end of Mount Murchison, at the Springs, and at the northern end of Scrope's Range, may be, and probably are, of Devonian age. They consist of hard, dense, thick bedded quartzites, similar in character to those at Mount Lambie, near Bathurst, showing slicken-sided joints, and as a rule lying at a high angle with the horizon. But the rocks at Wilcannia are of a different character. They consist of soft, yellowish, greyish, and whitish grits and sandstones frequently containing bands and pockets of kaolin, and lying as a rule at a very low angle of inclination. In fact, while one set of rocks shows abundant evidence of both metamorphism and disturbance, the other is remarkably free from signs of either. My conclusion in regard to these rocks is that they are probably of Upper Cretaceous age, and if this be correct it means that instead of the Cretaceous basin being cut off on the south by an east and west boundary through Wilcannia, there may be a deep channel somewhere between Woychugga Lake and Mount Manara, by which the artesian basin may have extended far to the southwards—possibly even under the Eocene beds of the Lower Darling, of the north-western portion of Victoria, and part of South Australia to the neigbourhood of Mount Gambier, where fresh water has long been known to escape as springs on the sea coast, as was first pointed out by the Rev. Tennison Woods ('Geological Observations in South Australia,' 1862.) It is quite possible, however, that the water may be derived from the Eocene beds themselves, and not from the underlying Cretaceous beds.

"In any case the probabilities of the artesian water-bearing beds extending southwards from Wilcannia, appear to be strengthened by the occurrence of Upper Cretaceous rocks (desert sandstone) at Bidura near Balranald, as reported by me in June last, and also by the fact that a deep channel has been proved to extend from Urisino (where two fine supplies of water have already been obtained) southwards along the west of the Paroo in the direction of Wilcannia—for I understand that several deep private bores were put down on Momba station. The deepest of these was 2,000 feet, but I believe that boring operations were discontinued before bed rock was reached.

"I propose during the coming year to make a geological inspection of the country along the southern course of the Darling, with the object of supplementing the information already obtained, but the only satisfactory way of settling this question is by boring. and I am of opinion that there is sufficient geological evidence to warrant the expense of a series of bores to the south of Wilcannia. In my journey northwards from Broken Hill the Upper Cretaceous rocks were first met with at Fowler's Gap to the north-east of Corona station. A good section of these beds is seen four miles west of Sandy Creek bore, and also about twelve to fifteen miles west of Bancanya bore, where they form the eastern escarpement of the Koko ranges. They consist of soft yellowish grey sandstones and grits often showing false bedding, and often stained by peroxide of iron. They are, in fact, in no respect distinguishable (lithologically) from the sandstones subsequently examined at Wilcannia. On the western flanks of the Koko range these sandstone beds are seen to lie unconformably on the upturned edges of slate rocks of probably upper Silurian age. The sandstones here dip to the east at a low angle (about ten degrees) but as they are followed eastwards the dip is seen to increase until at the eastern side of the range it attains an angle of forty-five degrees. It is unusual to find Upper Cretaceous rocks so highly inclined as this, but at least one instance of as high a dip as the above has been observed in Queensland by Mr. Rands (assistant Geologist).

"In many other localities as at Milparinka, Mount Poole, Mount Stuart, and in the Grey ranges, similar soft sandstones—but dipping as a rule at a very slight angle—are met with, and these frequently alternate with or are sometimes overlaid (conformably) by hard rocks, which, though somewhat of the nature of quartzites are perfectly distinct from the Devonian rocks, previously alluded to.

"Evidences of the agency of thermal springs are frequent in the Upper Cretaceous rocks. At the Peak (Mount Stuart Range) ancient thermal springs have left mounds of curiously banded limonite, showing that many of them contained ferruginous as well as silicious solutions; indeed the Upper Cretaceous rocks are characterised by the occurrence of considerable quantities of iron oxide.

' The whole of the country between the Waratta Ranges and the Queensland border is more or less overlaid by these Upper Cretaceous rocks, and they appear to extend for a good many miles to the east and west. In none of the country traversed by me were the Lower Cretaceous rocks (Rolling Downs formation of Queensland) to be seen outcropping at the surface, there being always a capping of either desert sandstone or of Pleistocene sands to hide them from view. But in the spoil heaps of many of the wells which have been put down by the pastoral lessees are to be seen the characteristic blue clays and sandy shales of the Lower Cretaceous formation."

The above are full extracts from the Government report of Mr. Pittman. I have given them thus fully as I consider they show a most valuable geological examination into the existence of the water-bearing rocks, enlarging very considerably upon the areas previously established, and—as in the examinations made by Mr. Jack, the Queensland Government Geologist, into the areas of that colony*—so far settle the question of the extent of the water-bearing areas of New South Wales.

Other and most important aspects of the subject are the source and probable quantity of the water that supplies these water-bearing rocks.†

Among all the numerous Government Reports and Papers published in the Australian colonies by explorers, Government officials, scientific societies, or university professors, previous to 1878, I can find no reference to or advocacy of a subterranean or artesian water supply. One of the first writers upon and advocates of that supply was the Government Astronomer of New South Wales (Mr. H. C. Russell), the foresight and scientific soundness of whose writings and opinions I have, I may say, always held in the highest estimation. In his "Rain and River Observations" for 1880 and 1881 Mr. Russell published some very valuable reports bearing upon this question. In his report of 1880 he writes:—" Since the rain measures of 1880 and the river measures for the same period are more complete than they have ever been before, it will be worth while to test

---

* See Section XI., "Queensland."
† See Section XVII., " Permanence of Supplies "

one by the other. I have before endeavoured to prove that the water passing down the Darling in an average year is only a very small portion of the rainfall, and is, in fact, very much less than must be available for that purpose after every allowance that can be made for evaporation and vegetation. For 1880 we have the means of testing this question by observations more complete than any which have previously been taken over the best part of the watershed of the Darling, that is, the western slopes of the Main Range, where, from the abundance of rivers and creeks, it is obvious that the rain-water readily runs off the soil. There are forty-five rain stations, and the mean rainfall derived from these is 20·74 inches; the area included is about 110,000 square miles. All the drainage from this country passes Bourke in the River Darling, and at this point a daily record of the height of the river is kept, and the mean result shows that the river has averaged throughout the year 6′ 8″ inches above the summer level. The width of the river at Bourke is 180 feet, and the velocity when in flood is rather less than one mile per hour. A few figures, which I need not give here, suffice to prove that one-quarter of an inch of rain over the watershed, or one-eightieth part only of the rainfall, represents all the water that passed Bourke during the whole year. When full allowance is made for the power of evaporation in a dry year, and for all other circumstances which might prevent the rain-waters reaching the rivers, it is certain that a very much greater portion than one-eightieth becomes running water. In such country as that under discussion common experience would give one-third of the rainfall as the available water, but for the sake of being on the safe side, we will assume that only one-tenth of the rainfall becomes running water, and it still represents a quantity sufficient to supply eight rivers like the Darling for the whole year. It therefore seems impossible to doubt that an unlimited supply of water passes away underground, more in fact, than would suffice to make the whole of the western districts a well watered country, and all that is wanted to make this supply available is a judicious use of the *boring rod.*"

In his Report for 1881 Mr. Russell further states:—' The evidence is conclusive, that the annual supply from rain finding its way into this great natural storehouse, is perfectly inexhaustible; it is also certain that as much must find its way out as in every year, under natural conditions, and the few wells that have been sunk prove that the outlet is so situated, that the water is under pressure in the reservoir, and will rise up to or above the surface when wells are sunk into it."

Mr. Russell spoke in 1881 with no uncertain voice. In a paper read before the Royal Society of New South Wales, in August, 1889, he says, "In a country like the interior of Australia, it is a surprise to many persons to find such an abundant supply of underground water, and many theories have been propounded about it. The high lands of New Guinea, and even more distant countries, have been mentioned as the source of the water, and when I pointed out just ten years since, the remarkable relations existing between the rainfall and rivers of the west, I was told by engineers and squatters, who knew all the country, that it was impossible that my statements could be true. It was most positively asserted, that no rain ever fell there that would wet the ground 18 inches deep, much less afford any for underground supplies ; with equal confidence it was asserted that what did get into the ground was all dried out again by evaporation ; and further, that the greater part of the Darling River basin was so flat, that water would not run upon it ; and that the rain therefore did not and could not, any portion of it, find its way into the river. To these and many other statements I felt that it was of no use making counter statements until facts should be collected that would give a fair basis for argument. I had stated the results of the first measures of rain and rivers, and they were so surprising, that few believed them, and I determined to wait until the results for several years should be available for discussion before speaking again, for the question at issue is of great scientific as well as practical importance. I may say in passing, that I do not think the question can be finally settled yet ; but the interval has sufficed to bring out important facts which I think should be published.

"First, in reference to the opinion, that no water reaches the Darling from the flat country, the following fact is sufficient to show that that view has been pressed somewhat too far, and will have to be modified so as to admit, that in heavy rains water does reach the Darling from it :—On the 21st January, 1885, a remarkable rain storm entered this colony in the N.W., not far from Milparinka, and travelled at the rate of about seven miles per hour, straight across country to the sea in an E.S.E. direction. On all the country round Wilcannia from ten to eleven inches of rain fell in forty hours. The river had been very low for months before, but sufficient water from this rain storm ran off the comparatively flat country to make a flood in the Darling at Wilcannia, that reached a maximum of 28 feet above summer level ; this flood did not subside to the old level until February 26, which was clear proof that the rain water not only filled the river, but continued to drain into it for several weeks. Certainly the water did not come down past Bourke, which, being in the margin of the storm, was but little affected by it ; and the river measures there showed that the only rise reached its maximum at 4 feet, and was all over in four days. There was no other possible way for it to come but off the country about Wilcannia where the rain storm passed over. It was obvious, therefore, that the opinion referred to must be taken with some reservation, for the instance just given shows that the Darling is in times of heavy rain, fed by the drainage of the country below Bourke, and this amounts to proof that at times it is fed by other parts of the flat country. With reference to the view, that the underground water comes from New Guinea or even more distant high land, that is, South America, it seems hardly necessary to answer it seriously ; it is quite certain, however, that even if the mountains of New Guinea could drain into our western plains the area of them is utterly insufficient to afford the supply.*

" It is impossible to say exactly how much is lost from the ground by evaporation. We could tell approximately how much was lost from the rivers, but the investigation of the loss from the soil is a very difficult matter, for reasons which might be easily explained. So long as the surface soil is wet the evaporation goes on from it rather faster than from water, but as soon as the soil dries down half an inch—which does not take long—the evaporation from it practically ceases, the layer of dry earth seems an effectual covering to the water below ; it is obvious then that the evaporation from soil does not go on at the rate which many persons suppose it does, and I am quite convinced that this source of loss does not seriously affect the underground supply ; once the water has sunk into the porous soil it is safe. This view of the effect of evaporation is borne out by the fact that the Murray is subject to similar temperature and wind to cause evaporation, and they do not dissipate all the water as some have supposed they do in the Darling country. This is a very strong argument against the view, so often expressed, that the small quantity of water in the Darling is due to the influence of evaporation upon the rainfall."† Mr. Russell goes on to say, " This remarkable condition affecting the discharge of the Darling river is one that calls for investigation, because if it be true, as I fully believe, and the observations prove it, then we

---

* See Mr. Jack's (Government Geologist, Queensland) corroborative opinion upon this.

† From careful experiments I have myself made, I can fully confirm these views of Mr. Russell upon the important question of evaporation.

must have a supply of underground water which is practically inexhaustible for pastoral purposes, and in addition to irrigate some of the land. The mean rainfall on the Darling river catchment for the past ten years (see 'Permanence of Supplies') has been 22.14 inches, and of this, as we have seen, only $1\frac{1}{2}$ per cent., or $= 0.33$ inches of rain passes Bourke in the river. If 25 per cent. of it, which is equal to 5.53 inches of rain, passed away in this river as it does in the Murray, there would be seventeen times as much water passing Bourke as actually does pass; but in addition to the water passing down the Murray we know that a certain amount of the Murray rainfall sinks into the ground to supply wells there; and hence 25 per cent. of the Murray rainfall does not represent all that is available from it. So much we find in the river, and some more, an unknown amount, is to be found in the soil. We should then be perfectly justified by the analogy of the two river basins, in assuming that the estimate just given of the amount of water which should pass Bourke is below the mark, not above it, and we ought therefore to have an underground water supply at least equal to seventeen times as much water as passes Bourke now, and this, or at least, a great part of it, should be useful for irrigation. That we do not find it in the Darling is, to my mind, proof that it passes away to underground drainage."

I think it is fortunate that those who were directly interested in the subterranean water supply of this country—Governments, Pastoralists, and Engineers—have had so valuable a coadjutor in their midst as the writer and exponent of such sound opinions as the above. I take it that Mr. Russell's observations of rainfall and river discharge were incidental to the immediate province of his department, but the application he made many years ago—as shown by his writings—of his observations with a view to the practical development of artesian water supply, and his more recent writings, are an invaluable addition to our knowledge of the laws regulating the subject.

In recent publications of Professor J. W. E. David, Sydney University, he traverses the ground formerly occupied by Mr. Russell. In "Notes on artesian water in New South Wales and Queensland" (read before the Royal Society of New South Wales, October 4th, 1893), Professor David has a notable passage as follows:—" During the whole of the rolling downs period large rivers, flowing from east to west, must have entered the margins of the Cretaceous Ocean and must have spread their gravels over extensive areas, and these rivers in New South

Wales at all events, continue to flow down to the present time, their channels of course being subject to oscillations of position and various modifications, as one geological period succeeded another. It cannot of course be affirmed that such rivers as the Dumaresq, the Gwyder, the Namoi, the Castlereagh, the Macquarie, and the Bogan were represented in Cretaceous times by rivers flowing in approximately the same latitudes as these modern rivers, but it can confidently be affirmed that in all probability several rivers, of which the above-mentioned are the modern equivalents, must have existed in Cretaceous time and have drained westward into the Cretaceous Ocean. Extensive gravels, must therefore, have been continuously in course of deposition from early Cretaceous time until the present, and unless the latitude of the main valleys has altered since Cretaceous time, the gravels belonging to successive geological periods must in places be superposed on one another, and so they would afford means of ingress for large bodies of water, which sink through the older gravels forming the channels of the modern rivers above mentioned. It is probable then, that a considerable amount of water finds its way into the artesian beds of the Rolling Downs formation, through percolating through ancient river gravels. The tunnel carried below the channel of the Macquarie river at Bathurst which tapped the water in the extensive gravel beds below the level of the river yielded large volumes of water without any evidence of the supply becoming diminished. In the opinion of the author there is a considerable quantity of water in these gravels which underlie the channels of the modern rivers, and perhaps as much (if not more) water drains through them into the delta gravels formed by the Cretaceous rivers, and so into the vast sand beds of the Cretaceous formation further west, as drains into all the outcropping edges of the Cretaceous porous beds occupying the areas intermediate between the positions of the former estuaries of the Cretaceous rivers."

In the same paper (October 4th, 1893). by Professor David, is the following:—" In Queensland Mr. R. L. Jack, F.G.S., the Government Geologist, and Mr. J. B. Henderson, the Government Hydraulic Engineer, had with remarkable foresight and courage spoken in no uncertain tones as to the comparatively limited extent of the artesian water supply, and he would like once more to emphasise the fact that in New South Wales also the supply was limited, so that possibly by the time that the quantity of water drawn from the artesian wells had been increased ten or twenty-fold it would be found that the demand had overtaken the supply—that the annual outflow had equalled the annual intake of rain water into the artesian water beds."

In the last paper read by Mr. Jack (that before the Australasian Association for the Advancement of Science, Brisbane, 16th January, 1895) is the following:—*" The amount of water contributed to the water-bearing strata of the Lower Cretaceous formation every wet season by such rivers as the Darling is so great, and consequently the amount of leakage into the sea is so great, that the quantity abstracted by the artesian wells, large as it is, and even if it were ten times greater is insignificant by comparison. Finally, as the leakage into the sea is so vast and is entirely beyond human control, the draught on our underground supply by artesian wells is not worth controlling. I make no apology for the fact that my views on this important question are not those which I held twelve months ago." In view of this very decided position now taken by Mr. Jack as to the volume and permanence of supplies, one cannot help regretting that so prominent an authority as Professor David should have spoken in an adverse manner upon the subject.

A *resumé* of the position of New South Wales as regards its underground water supplies may be stated as follows :—

1. There is a very large area (62,000 square miles) of water-bearing country.

2. Most of the rainfall, as shown by that of the River Darling catchment area, sinks into the ground and acts as feeder to the intake waters at the outcrop of the water-bearing rocks. Of this I cannot conceive any doubt whatever. If we accept the figures of Mr. Russell, Government Astronomer, and there is every reason to feel confidence in his measurements and deductions, we get a surprising result, viz., that in all probability sixteen times the water that the Darling carries away passes underground. The measurements of the rainfall and river flow extended over ten years, and show that this one river alone disposed of only $1\frac{1}{2}$ per cent. of the rainfall on its catchment areas.

3. That this underground water is being constantly tapped by artesian borings, and that its development and utilisation on a very extensive scale may be taken to be assured in the near future.

---

* See Section XI., " Queensland."

Artesian Wells.

The following is a List of the Government Bores completed and in progress :—

Description.	Depth. feet.	Remarks.
Arumpo, Euston Pooncarie	1020	Bore determined
8-Mile, Waanaaring Road	230	
Tolano, Ivanhoe, Menindie	1603	Pumping supply at 800 ft.
Gidgea Camp, Bourke, Hungerford	882	14,000 galls. per diem, good fresh water
Green Camp, Nyngan	1506	Good pumping supply, 800 feet from surface
Pack Saddle, Silverton, Cobham	1942	
Hay, Hay		Bore stopped, good pumping, supply at 800 feet
Cuttaburra, Bourke, Waanaaring	1723	Bore completed, bed rock reached at this depth, good pumping supply at 850 feet
Pera, Bourke, Waanaaring	1154	Completed, 700,000 galls. per diem of good fresh water
Dolgelly, Moree to Bogabilla	2275	
Bancannia, Silverton, Cobham	1357	Work suspended
Poison Point	1334	
Nevertire, Nevertire	2116	
Bendermere, Brewarrina	1725	Bore determined, bed rock reached. This bore is outside recognised artesian water bearing formation
Moree, Moree	2320	350,000 galls. per diem, boring proceeding
Gilgandra	2090	
44-Mile (Clifton), Milparinka, Waanaaring	1638	2,000,000 galls. per diem, temperature 138° Fah.
Kulkyne, Bourke, Waanaaring	1498	
Osaca (76-Mile)		
Waanaaring, Milparinka	1646	350,000 galls. per diem
Brigalow, Culgoa, Lidknapper	2150	300,000 gals. Work proceeding
Dolmoreve Well	1240	Bore determined
Waanaaring, Waanaaring	284	Work proceeding
Tineroo	650	
Berrawinnia, Hungerford	885	Bore determined
Quarry Reserve, Bourke	785	
Marra Creek, viâ Coolabah	851	
Gil Gil, Moree, Boggabilla	1802	100,000 galls. per diem
Warratta Milparinka	25	

KERRIBREE, BOURKE DISTRICT, NEW SOUTH WALES.
Depth, 1,340 feet; Flow, 1,700,000 gallons daily; Temp, 100 deg. Fah.

SECTION XI.

QUEENSLAND.

QUEENSLAND must be taken, in a material sense, into the category of very fortunate communities. Nature has not only provided the colony with exceptionally rich mineral resources and soils and climates capable of providing the widest range of vegetable products in the greatest perfection, but her wealth of underground waters—her liquid assets—lying as they do under a greater part of the colony are, there is every reason to believe, enormous, and so long as the rain falls will in all probability prove inexhaustible.

The assertion that the underground water supply of Queensland will be of infinitely greater value to the country than all the gold mines that have yet been discovered may startle many who have not yet studied the subject. Yet such is an indubitable fact. The discovery of artesian water has already saved stock to the value of many thousands of pounds, and when the immense water-bearing areas hitherto subject to drought have been further tapped by boring the saving in future years will amount to millions more, and at the same time make agricultural pursuits profitable in districts where scantiness of rainfall renders them too precarious to be thought of.

The presence and plentifulness of artesian water depends on the rainfall in the higher regions, on the lower altitude of the bore site, and on the permeable character of the rocks between. The territory of Queensland is in an exceptionally favourable position for the fulfilment of these conditions. The rainfall on which most of its artesian water depends is that caught on the western slopes of the Dividing Range, from which almost the entire country to the border trends downward. It is not because the rainfall is so scant that the Western river system runs dry half the year, but because so much of it sinks into the ground. Then the formation of rock most favourable for the subterranean carriage of water—the Lower Cretaceous—is the prevailing formation over the whole of the wide-stretching Western Downs, embracing an area, as calculated by Mr. Jack, Government Geologist, of 376,832 square miles—equal to over 56 per cent. of the total superficial area of the colony. A fact like this opens up vast possibilities for the future. With the exception of the Dakota basin, in America, the artesian basin of Queensland is the largest yet discovered in the world.

The first sub-artesian water obtained by boring in Queensland was by Government in 1882. This was at the 14-mile Dam, near Cunnamulla.

The first artesian water obtained was at Back Creek, 20 miles from Barcaldine, on the Queensland Central Railway, but this did not amount to much, the water rising only three feet or four feet above the surface. The bore was sunk at a very low spot, and in all probability was not from the same source as that obtained subsequently further west.

The first artesian bore started in Queensland was that at Blackall (Government) in January, 1886, but it was abandoned waterless at a depth of about 1000 feet, the cable system which had been used in drilling it proving unsuitable. Work was recommenced some time after the completion of the Barcaldine bore next described, and a depth of 1667 feet reached, when a splendid flow of 300,000 gallons per diem was obtained.

It is generally held that the Barcaldine Government bore can fairly claim to have been the first important artesian supply tapped in Queensland. The tools to sink it with were landed from Canada in July, 1887. The machinery was made principally in Sydney. The bore was commenced on the 18th November, 1887, and water was struck on the 28th December following. It is needless to say the work was watched with the keenest anxiety, and when after six weeks' work water was struck at a depth of 691 feet there was great rejoicing. When the Barcaldine water baptised the Western Plains it was felt that a new lease of life had been secured, and that the drought had been robbed of some of its terrors. The flow of water at Barcaldine bore is 175,000 gallons per diem. After the success of this bore the plant was moved further along the railway extension, and water was struck at a depth of 978 feet. During the time the Barcaldine bore was being put down a plant had been moved to Blackall, and the bore there deepened to 1663 feet. This bore was completed shortly after the railway bore— known as the 21-Mile—had struck water, and a fine flow of 300,000 gallons per diem was obtained. This was about April, 1888. The plant was then moved to Tambo, where water was obtained at a depth of 1002 feet; while the Government plant which had been at work at the " 21-Mile " was removed to a site on the railway line on Wellshot Run. It, however, sustained a series of accidents, and the site was abandoned in June, 1889, and the plant moved to Winton, where a bore was put down to 1125 feet. The next bore was made by the Aramac Divisional Board, on the Aramac-Barcaldine Road, in February, 1890.

CAMBRIDGE DOWNS, No. 3, QUEENSLAND.
Depth, 597 feet; 500,000 gallons per diem; Temp., 98 deg. Fah.

Work was being carried out at this time on Saltern Creek Station. The first bore struck water at a depth of 570 feet, but was carried down to 1130 feet, when a supply of 175,000 gallons per diem was secured. This bore was the first to settle the then vexed question as to whether or not a second and greater volume of water existed below the proved first supply. The answer was in the affirmative, and boring operations were then pushed vigorously and earnestly ahead. Two more bores were successfully sunk on the same station yielding 4,350,000 gallons of water daily. Previous to this a second bore had been made at Barcaldine, and this also proving successful the fact of artesian water existing all along the Western Downs within a certain distance of the edge of the desert was established.

Early in 1891 water was struck at a depth of 2700 feet on Darr River Downs, and this bore being carried to a depth of 3630 feet a large supply—500,000 gallons—of water was struck. By this time water had also been struck on Warenda Station, near Boulia, and this fully established the fact that artesian water was to be got all over the Western Downs, depth being the only consideration. Boring was then pushed into the north-west country, and was most successful on the Flinders, at Richmond Downs, Cambridge Downs, Marathon, Saxby Downs, and other stations. Good results were also obtained at Landsborough Downs, Barenga, and Afton Downs, and is being carried on at these stations and also at Katandra. Probably the most successful stations, when result and cost are taken together, are Coreena, Aramac and Stainburn, the two last being to quote the words of a borer, "able to run creeks on their stations at pleasure."

While work was so satisfactorily progressing in the Central and north-west districts many bores were under weigh in the south-west, and here amazing results were obtained. At Burrandilla two startling overflows were secured, one of 4,000,000 and the other of 2,500,000 gallons daily. No. 2 bore on Charlotte Plains yielded, at a depth of 1848 feet, 4,000,000 gallons; Coreena bores, 2 and 5, yielded respectively 1,500,000 and 1,000,000 gallons; while on Tinnenburra seven bores threw out 8,000,000 gallons of fine water daily; whilst Boatman bore, No. 1, discharges 4,200,000 gallons in the same time. And so the work has gone on until it is computed that there are at the present time nearly 350 private bores in the colony of Queensland, from which over one thousand million gallons of water are flowing daily.

There are only two instances recorded where artesian water has been found on the eastern slopes—between the coast and the ranges—and those are at Laidley, where a very small supply was met with at a depth of 2512 feet, and at the Racecourse bore, near Brisbane, where a small supply of 8228 gallons per diem was tapped. This water proved to be unfit for domestic purposes, and the bore was abandoned.

In Appendix A will be found a list (from the last Government Report, November 1st, 1894) of the Government and private bores in the Colony, additions being made and the list brought up to date by the Author.

The total number of feet bored in Queensland artesian wells known to me at this date is about 486,000, equal to about 92 miles of boring.

Boring plants were at first imported from Canada and from Germany and admitted into the Colony duty free, but this has now been wisely stopped. Complete plants are now supplied by Brisbane, Rockhampton and Townsville manufacturing firms.

Some of the recent "finds" of water are those on Dillalah Station, near Charleville. A flow of water was struck of 2,500,000 gallons daily at a depth of 1990 feet. The bore is situated on a ridge at the junction of three paddocks; the water conservations in each have been filled for miles around the bore, and this was the means of keeping the stock alive during recent droughts.

A still more important "find" is that at Boatman Station, Bullon, where a wonderful flow of 4,200,000 gallons per diem has been struck at a depth of 1500 feet. In the enormous output of perfectly good water at a comparatively moderate depth this work will rank amongst the most successful bores of Australia, or, in fact, of the world.

Since the completion of the bore the water is being utilised in a highly satisfactory manner by the station lessee, and an account of this collateral work may be offered as a picture of the realisation so long advocated and looked for by well-boring engineers in this country, and as the best possible evidence of the possibilities in the development of the "liquid assets"—the subterranean water—we have with us.

One-half of the vast stream of over four million gallons of daily discharge has filled eleven dams on the Nebine and the Creek also for a distance of twenty miles to the station boundary, and the other half has been carried by drains for thirty miles back. The drains are four feet wide and nine inches deep, and are to be further extended. The sheep water at them without difficulty and without damage to them. This system waters eight paddocks in what was formerly dry country. It is the intention to excavate small tanks in which to conserve the water, and only turn the bore water on when they require replenishing. If No. 2 bore, now in hand, be successful it is proposed, by means of a blind creek and its ramifications, to water about 400 square miles of what is at present unavailable country, and which will increase the carrying capacity of the run by 100,000 sheep. No. 2 bore, on the same run, is now in progress.

At Currawinga Station, west of Paroo, at a depth of only 250 feet, a nice supply of 84,000 gallons per diem of artesian water has been obtained. The bore was put down by the station hands with a drill or jumper made on the station. *(Vide* section, " Shallow Well-boring.") The locality is one where the supply will be of immense benefit, while the unusual shallowness of the bore is most remarkable, all water west of the Paroo being found previously in the region of 2,000 feet, the deduction from which is unavoidable, *i.e.*, that artesian water can be obtained by boring in the areas of the water-bearing rocks if the drill goes deep enough, and that it may also be found at comparatively shallow depths.

The following gives the results of the recent (January, 1895) investigations of the Government Geologist (Mr. R. L. Jack, F.G.S.) into the artesian water-bearing areas and formations of Queensland :—*

" It is now well known that all our artesian water, with trifling exceptions, occurs in the Rolling Downs or Lower Cretaceous formation. Over this formation the Upper Cretaceous or Desert Sandstone lies uncomformably. The latter must have covered an area of at least 500,000 square miles, but has now been reduced by denudation to isolated tablelands. In mapping the eastern limit of the Lower Cretaceous formation we find at the base there is a series of soft grey, very friable sandstones, grits, and conglomerates. This sandstone absorbs water with avidity. The rock is, moreover, so destitute of cement—or it may be that the cement is so soluble—that a lump of it on being saturated with water falls away to a heap of sand. We can, therefore, understand how underground, where such strata are saturated with water, they may be correctly described by the drillers as ' sand ' instead of ' sandstone.' To this rock we gave the distinctive name, the Blythesdale braystone, as it is well developed at Blythesdale, near Roma. This sandstone is first met with

---

* Paper read before the Australasian Association for the Advancement of Science, Brisbane Meeting, 1895.

near the New South Wales border—at Whetstone, on the Macintyre River. The exigencies of travel took us down the Macintyre to Goondiwindi and up the Weir River, where nothing but alluvial soil was met with till we reached Tarewinnaba, on the Weir. At this point the braystone is just seen, and is at once covered by the Desert Sandstone. A cake of Desert Sandstone extends from this point north, east, and west. The eastern and northern margins of this tableland were traced, and it was found that the Desert Sandstone overlaps the Ipswich coal measures, which we exposed in several places where the Condamine has cut through the overlying sandstone. On the north-western margin of this tableland, near Surat, the Desert Sandstone directly overlies the upper or shaly beds which give rise to the Rolling Downs, and which form by far the greater proportion of the strata of the Lower Cretaceous formation, so that the Blythesdale braystone may be assumed to crop up beneath the Desert Sandstone somewhere between Dogwood Creek and Wiconbilla Creek. At Bendemere, on Yeulba Creek, the Desert Sandstone having been denuded by the stream, the upper shaly Rolling Downs beds are directly in contact with strata understood to belong to the Ipswich coal measures, and the non-appearance of the Blythesdale beds is accounted for by a 'fault.' A few miles to the west the Blythesdale braystones are met with in great force, being succeeded to the north by the older (supposed) Ipswich coal measures, and to the south by the newer shaley members of the Lower Cretaceous formation. As the braystone is traced north-westward by Blythesdale, Roma, Taboonbay, Donnybrook, Hogauthulla, and the heads of the Warrego, its south-western edge merges into the Downs formed by the shaley members of the Lower Cretaceous, and to the north-east it disappears under the Desert Sandstone (which is itself succeeded by bedded basaltic lavas). Nothing is more certain than that no member of the Lower Cretaceous formation appears on the north-eastern side of the Desert Sandstone and basalt tableland which form the divide between the waters draining southward into the Great Australian Bight and those draining eastward into the Pacific. In some places, however, as at the head of the Warrego on the Bight side, and at the head of the Nogoa on the Pacific side, of the almost imperceptible watershed, the denudation of the Desert Sandstone (which apparently has a very uneven bottom) exposes larger or smaller areas to the Blythesdale braystone.

"On the divide between the head waters of the Warrego and Barcoo the Desert Sandstone remains as a tongue which covers

the whole of the outcrop of the Blythesdale braystones, and on its western margin directly overlies the shaley members of the Lower Cretaceous formation. A glimpse of the braystones is seen at the head of Birkhead Creek, covered to the north and east by the Desert Sandstone, and succeeded to the south-west by the shaley beds of the Rolling Downs. Further north, on the head waters of the Barcoo and the Thompson, the Desert Sandstone extends as a comparatively narrow tongue as far west as Barcaldine. From beneath this formation the Blythesdale braystones emerge at a point on the heads of Aramac Creek, about thirty-five miles north-west of Jericho, and have been followed northward through nearly two degrees of latitude, until it is again covered by the Desert Sandstone in the vicinity of Corinda Station. The western edge of the Desert Sandstone rests directly on the Lower Cretaceous shales from this point as far as Hughenden, and the Desert Sandstone forms an unbroken escarpment of about eighty miles in length, and entirely conceals the Blythesdale braystones from view.

"In the region where an attempt has been made to map the whole area covered by the outcrop of the Blythesdale braystone, namely, from near Chadford, on a head of Yeulba Creek, to the Warrego, the outcrop forms a belt varying in width from five to twenty-five miles. The invariable position of this belt on the outer edge of the Lower Cretaceous area leaves no room for any reasonable doubt that it is composed of the beds lying at the base of the formation; but apparent dips have proved confusing and unreliable, so that I have no confidence in any estimate of the total thickness of strata represented by the outcrop. I incline to the belief that on the whole the angle of dip is very low—perhaps in many instances not more than the fall of the ground. However, an angle of one degree, or 1 in 57, would give a thickness of 400 feet for an outcrop five miles wide, and from the evidence afforded by numerous bores I am prepared to admit that this thickness is by no means over-estimated.

"Hitherto it has been convenient to speak of the series of beds designated the Blythesdale braystones as being of similar composition throughout. This, however, is not the case, as the braystone of normal composition is 'parted' in places by beds of sandy shale and calcareous sandstone. We may imagine coarse, sandy and gravelly sediments brought down to the margin of a shallow Lower Cretaceous sea by numerous tributing rivers, and spread out along the shore and out to sea by the action of waves and currents. Such material could only travel along the bottom, being too coarse and of too high specific gravity to remain in suspension, as a general thing, far from the mouths of rivers. The influence of wave-action, at least, would cease when the sea attained even a moderate depth. I am inclined to believe that the sea in which the Blythesdale braystone was deposited and distributed—namely, the sea which divided the Australian Continent from north to south into two islands—was very shallow throughout, and may have been swept from end to end by currents sufficiently powerful materially to aid wave-action in the distribution of the sand and gravel. Otherwise I do not know how to account for the widespread distribution of the sand and gravel which is evidenced by the artesian wells. I imagine further that the later but continuous period to which the argillaceous Lower Cretaceous deposits belong was one in which a marked subsidence of the interior of this sea took place. The wide distribution of the Blythesdale sands has a parallel in a newer geological period in the enormous extent of the Desert Sandstone. Unless I had seen it with my own eyes, I should have had a difficulty in conceiving that heavy gritty sand could be so widely distributed as it is in the latter case; but having seen it I can believe that the Blythesdale sands may be equally extensive. It may be assumed that the Blythesdale and Desert Sandstone periods were both characterised by heavy rainfall, producing rapid denudation of the land, and possibly by prevalent rough weather at sea causing violent wave action. At the same time the intercalation of some argillaceous sediments among the Blythesdale braystones need cause no surprise, as they would be the natural result of occasional spells of dry weather on land or comparative calms at sea. The wonder is that there should be so few. It is not likely that the intercalated argillaceous sediments are continuous over the whole area occupied by the Lower Cretaceous formation. They are more likely to be lenticular at the margin, and to spread out and thicken towards the interior of the area. The Blythesdale braystone, therefore, although it may locally be split up into two or more beds by the intercalation of comparatively impermeable strata, I believe to be practically a continuous deposit. It has already been mentioned that the Blythesdale braystone is, over a large area, covered by the Desert Sandstone. Taking this into account, and also the fact that the actual breadth of the outcrop, where it is not covered by the Desert Sandstone, has not actually been mapped, we are still uncertain of the total area of the outcrop. I think, however, we are safe in assuming an average breadth of at least five miles. In his annual report for the year 1892 Mr. Henderson made

SESBANIA, QUEENSLAND.

what I then thought a very liberal allowance when he based a calculation of the area of gathering ground on the assumption that 'the aggregate breadth of the outcropping edges' was one-eighth of a mile. In the light of the recent investigation, I feel tolerably confident that the breadth of the basal beds (the Blythesdale braystone) is at least forty times as much, not to speak of beds of similar composition on higher horizons, which makes a material difference in the conditions of the problem. Assuming that the total length of the ribbon representing the outcrop in Queensland of the Blythesdale braystone is 1000 miles, and its average breadth five miles, this outcrop alone would give a gathering ground or intake of 5000 square miles. An average of thirteen meteorological stations along the line of outcrop (taken from the map issued with Mr. Henderson's last report) gives roughly for this area a mean rainfall of 27 inches, which is considerably greater than that of the Downs country.

" The computive altitude of the outcrop of the Blythesdale braystones and the western country where the artesian wells are situated is an all-important factor in the calculation. The braystones attain their highest observed altitude of 1700 feet above the sea level at Forest Vale, on the Maranoa, and the altitude gradually falls to 800 feet at the New South Wales border, and probably is almost at the sea level near the Gulf of Carpentaria. I cannot speak with confidence of the extension of the lowest beds of the Lower Cretaceous formation in New South Wales and Victoria, but as the cliffs on the coast near Coorong are supposed to be Tertiary the lowest beds of the Lower Cretaceous, if present, must be beneath the sea level at the Great Australian Bight.

" The outcrop of the Blythesdale braystones is crossed by several large streams—Blyth's, Bungil, Bungewoogorai and Amby Creeks, the Maranoa River, Hoganthulla Creek, the Warrego River, Birkhead Creek, and the eastern tributaries of the Thomson River. All these streams run only for a small portion of the year, but while they run a rock of the bibulous nature of the Blythesdale braystone must be absorbing water greedily and the water must not only spread laterally but also fill up as much of the underground portion of the stratum or strata as had been emptied by leakage. Two kinds of leakage might affect the bibulous beds at the base of the Lower Cretaceous formation in a sufficient degree to be worth consideration for the present purpose. Suppose the beds to dip seaward and beneath the sea, and either to rise to the ocean bed or to dip at a lower angle than the slope of the sea bed there would be a leakage into the sea. And again, suppose (what we believe to be actually the case) the outcrop of the beds to occur at gradually lower levels till it attains the sea level, there would be a leakage in the form of springs or into river beds all along the line. In either case the leakage, however compact the beds might be, would not cease till the water-level in the beds was reduced to the level of the sea, unless the head of water were from time to time replenished."

Mr. Jack then quotes Mr. Russell's (Government Astronomer, New South Wales) measurement of the flow of the River Darling (see "Permanence of Supplies") and proceeds:—" Mr. Russell's figures, if applicable to Queensland, prove more than the absorption of the water by porus strata. If they are reliable (and I see no reason to doubt their accuracy), and again, if they are applicable to Queensland (which is very likely, considering the exceptionally bibulous character of the Blythesdale braystone) the bibulous rocks at the base of the Lower Cretaceous must annually, or at least every wet season, absorb an amount of water which would severely tax my arithmetic. And if the bibulous rock absorb such an amount of water, they must first have been drained of water to a like extent! How would such a drainage be effected? Certainly not by the bores, or there would have been a difference in the character of the rivers of the district before and after the commencement of boring; and the output of the bores, great as it is, is after all a mere bagatelle compared to the amount which would be "lost" by the rivers of Queensland, supposing them to behave like the Darling. Certainly the drainage of the Blythesdale braystone is not effected by the bores. The only conceivable agent capable of effecting it is the leakage in the sea bottom. Thus, although we have no direct proof the latter, I hold that it is proven if once it be shown that the Queensland rivers suffer a loss of water comparable to that of the Darling. If the foregoing conjectures as to the drainage of the upper portion of the Blythesdale braystones and their periodical replenishment by rivers in the wet season be correct, it is possible to conceive of a long drought reducing the water to the sea-level in such of the strata as are open at one end to the sea. If the strata can imbibe as much as the parallel case of the Darling would lead us to infer, they must lose by leakage into the sea nearly as much as they imbibe, and the amount drained from them by the bores will be insignificant by comparison. If the water-bearing strata communicate with the sea the water can never fall below the sea level, but if it stood long at that level it would become salt. It has already been

RICHMOND DOWNS No. 1, QUEENSLAND.
680 FEET DEEP; 1,000,000 GALLONS PER DIEM.

mentioned that the Desert Sandstone covers a large area of the outcrop of the Blythesdale beds. Although the former is less bibulous than the latter it is still very absorbent. Mr. Tolson's remarks about 10 in. of rain having fallen on this formation without making the streams run are very much to the point, and I can confirm his observations from a large experience of Desert Sandstone country. Now the Desert Sandstone, thus saturated with water, lies like a full sponge on the top of the outcrop of the Blythesdale braystone, and must tend to equalize the supply by feeding the latter long after the rivers have ceased to run."

Mr. Jack then treats of the probability of the lowest beds of the Cretaceous receiving contributions of water from older formations, which it succeeds unconformably, and continues: "I have spoken so much of the Blythesdale braystone that I may have created the impression that they are the only water-bearing beds of the Lower Cretaceous formation. Their importance warrants the attention they have received, but there are water-bearing beds on higher horizons. Many of the bores have struck two or more supplies of water. Sometimes the first supply does not rise to the surface. A second may reach the surface and flow over, while a third may considerably increase the supply. The feeble supply of the upper beds may be explained by the fact that they do not crop out at high altitudes. As a rule the higher beds must be more limited in extent, and must draw their supplies for the most part from local sources. In other words they crop up where the rainfall is least, and the outcrop has less chance of being crossed by streams running long enough to replenish them."

In conclusion Mr. Jack says: "The amount of water contributed to the water-bearing strata of the Lower Cretaceous formation every wet season by such rivers as the Darling is so great, and consequently the amount of leakage into the sea is so great, that the quantity abstracted by the artesian wells, large as it is, and even if it were ten times greater is insignificant by comparison. Finally, as the leakage into the sea is so vast, and is entirely beyond human control, the draught on our underground supply made by artesian wells is not worth controlling. I make no apology for the fact that my views in this important question are not those which I held twelve months ago. My colleague, Mr. Maitland, and I went out into the field with the object of laying a foundation of facts for the guidance of ourselves and others who may wish to rest their theories on solid ground. It is for others to judge whether we have succeeded."

The above account I may say fully confirms my own observations and notes taken during my practice as a civil engineer during the last 13 years in Southern, Central and Northern Queensland, and previously in New South Wales and Victoria. I have given the greater part of Mr. Jack's account, as he being Government Geologist is naturally looked upon as an important authority on artesian water supply, although it is to be regretted that more prominence was not given in his earlier reports to that supply, or that former Governments did not earlier move in the matter.

My notes cover careful examinations of outcrops of the artesian rocks, both in travelling over a very large part of the colony, with the advantage of being located in different districts for some months at a time specially engaged in the business of underground water supply in all its bearings, both geological and practical. Those notes, I need scarcely remark, are fully utilised in this book, especially in the Section "Permanence of Supplies." I think it is to be regretted that Mr. Jack in his valuable paper did not give collateral evidence of the soundness of his views which may be found in the American Government and other Reports and Papers. The only reference made by Mr. Jack to the opinions and views of our scientific and highly informed coadjutors in America, is in combating a theory of "rock pressure" as partly accounting for the outflow of artesian water formulated by Mr. Robert Hay in the American Geologist (Vol. V. page 300.)

There are further portions of Mr. Jack's paper which are, I think, open to discussion, viz. :—(1) "An idea prevails in some quarters that every successful artesian bore is successful because it has struck the channel of an underground river; but against this theory some fatal objections may be raised. A river ceases to be a river when it enters the sea. Now, the Blythesdale beds are marine, as their outcrop (the nearest part to the old coast-line) contains marine fossils, as do the argillaceous strata above them; and we cannot imagine a river bed meandering through a sea-bottom. Besides the wide area over which the water-bearing beds have already been met with in bores, renders the idea that they are all river beds in the last degree improbable."*

---

* My belief is that artesian water is movable along "water sheets" or travels in channels or "underground reservoirs," and that this is evident from the fact of the water upon being tapped rising to the surface so freely and with, in most cases, the great force and great velocity and volume that it does, which conditions could not exist if the water issued directly out of the water-bearing rock instead of flowing in a channel, formed by its passage, out of that rock.

Mr. Jack's description of the water-bearing strata—the Blythesdale braystone—is that of a highly bibulous soft rock which crumbles by the action of water "that a lump of it on being saturated with water falls away to a heap of sand." In his definition of artesian water Mr. Jacks says :—" The stratum (the braystone) is simply a pipe of more or less sectional area which the bore converts into an inverted syphon. In a clear syphon the water would rise with all the force due to the pressure of the head of water; but a syphon closely packed with sand is not only a syphon but also a filter, and in proportion to the openness or compactness of the sand the flow will be strong or feeble." How the great volume and force and consequent velocity of the flow of artesian water upon being tapped is to be reconciled with a closely packed syphon I cannot conceive : the hydraulic pressure due to the head of water—that above the level of the surface at the bore—would certainly force the sand up the bore as mud rises to the surface from mud springs. In boring in the United States in my own experience, live fish have been forced up to the surface, the nearest river which similar fish inhabited crossing the outcrop or intake area of the water-bearing strata, as similar strata according to Mr. Jack's paper cross similar rivers in Queensland. If the inlet area at the bottom of the bores could be exposed to view it would I believe be found to consist of an open channel or reservoir, for without such a condition it is difficult to account for the free passage of such an enormous volume of water at so high a velocity as the bores discharge. The channel or reservoir theory is I submit, more consistent with the speed and force of flow. "Falling of the tools " is a common occurrence in well boring when the chisel—or working tools—sink suddenly on the water-bearing strata being reached, passing, as I believe, through an open space below the superincumbent impermeable rock and into the soft sandstone—the braystone of Mr. Jack's nomenclature.

Among many facts ascertained by the aid of boring it is proved that in strata of different ages and composition there are passages by which the subterranean water circulates. Thus, at St. Ouen, France,* five distinct sheets of water were intersected in an artesian well, and from each of these a supply was obtained. In the third water-bearing stratum a cavity had formed in which the boring tool fell suddenly and the water ascended in great volume. The same falling of the tools has taken place in Great Britain. In one case a deep bore was

---

* Degousées great work on artesian wells, "Guide du Soudeur Artesién." Paris, 1847; or Public Library, Melbourne.

sunk, when the water suddenly brought up a quantity of fine sand with vegetable matter and fresh-water shells, branches of a thorn several inches long; marsh plants and some of their roots were also brought up in such a state of preservation that showed they had not remained more than three or four months in the water. In this case it is evident that water had penetrated to great depths, not simply by filtering through a porous mass—for then it would have left behind the shells and fragments—but by flowing through open channels in the earth.

"Another idea," says Mr. Jack, "is that there are in the bowels of the earth vast reservoirs of water," and goes on to say that "I can imagine the roof of such a reservoir tumbling in," and that "my belief is, however, that such underground reservoirs are not common objects in nature." In regard to this belief it must be borne in mind that mining for petroleum shows the oil to be, as compared to water, a thick, viscid liquid, and yet the flow from deep bores made into it, both in volume, velocity and force of discharge, is similar to the discharge of artesian water. This oil is known to exist in reservoirs between non porous rocks. If it had to produce itself direct from the oil-bearing strata the velocity or volume of the flow is inconceivable, and there is no record of "tumbling in" of the roof either in Petrolia or Artesia. If most of the artesian water in its flow to lowest levels had not worn channels and formed underground rivers in its course the velocity of the water at its inlet at the bottom of the tubing or bore, on being tapped, would wear away the soft water-bearing strata, and the result would inevitably be a very copious and prolonged discharge of sand at the mouth of the bore. The fact that sand is not discharged in any marked degree is therefore, I think, conclusive proof that the theory of underground rivers in connection with artesian water supply is a sound one.

The flow of the bulk of the artesian water in Queensland, as in the other colonies, is, I believe, from the north-east to the south-west, and its ultimate discharge takes place in the bed of the ocean (the mouth of the main channel is supposed to exist near Coorong, on the South Australian coast, or somewhere in the Great Australian Bight). If this belief be a sound one, it follows that the flow must be on the basis of a river system. Long, long ago, before even the present flora and fauna of the country had begun to exist, there was a great chain of mountains which ran off at right angles from the Dividing Range. Its western limit was situated somewhere in South Australia. These mountains were of immense altitude, and the land lying

TOORAK No. 2, QUEENSLAND.
Depth, 1,550 feet; 860,000 gallons per diem; Temp., 140 deg. Fah.

between their feet and the sea presented a uniformly steep gradient. Years and years of rain, falling on their high tops, carried down the loose soil into the channels of the first rivers. By this process of erosion and denudation the beds of the rivers became elevated above the surrounding country, and when the proper time came the water was diverted from its course, and found a new channel. And this process was repeated over and over again, till the mountains sank and the plains rose, and the solid body of the Australian Continent was honeycombed with ancient river beds; and when the rain falls on the surface of the land it finds its way to these channels, which are maintained partially open by reason of the ancient *debris* which marks their course.

Sir Thomas M'Ilwraith, when Colonial Treasurer, in making his Financial Statement to Parliament, spoke as follows :— " Artesian water has been discovered under a wide area of the colony of Queensland. The Government initiative test bores— all, with one exception, successful ones—have been purposely made at great distances apart in order to test all the outlying portions of the Colony. Private bores have been made intermediately, and I think it is fair to assume that the country between these successful bores is also water-bearing. If that be so it may be concluded that water will be found in greater or lesser quantities, but in all in profitable quantities, in the following districts :—About 15,000 square miles of the south-eastern portion of the district of Burke; an isolated area of about 1300 square miles in Gregory North; about 20,000 square miles along the eastern boundary of Mitchell, one part of the area entirely crossing the ' district '; about 27,000 square miles along the eastern and southern boundary of the Warrego district; and about 25,000 square miles on the western and southern boundary of the Maranoa district; total, about 88,300 square miles. In other words, the area may be distributed as follows :—From Hungerford to Mungindi, on the New South Wales border; thence *via* St. George and Muckadilla to Charleville ; thence by a wide strip of country *via* Tambo, Blackall, Barcaldine, Aramac, and Muttaburra to Hughenden ; thence down the Flinders to the west of Manfred Downs, and as far south as Mackinlay. Large volumes of overflowing artesian water have also been tapped near Boulia, an isolated area in that locality. When we consider that this 88,300 square miles (the proved water-bearing area) comprises some of the best land in Australia, and that artesian water in the enormous quantities found puts cultivation by irrigation in the far west among the probabilities, I think we

F

have good cause for looking forward with much better hope to closer settlement in the near future within our big western territory."

Marked success has attended the efforts of the Queensland Water Supply Department under the Chief Engineer (Mr. J. B. Henderson, C.E.), to procure a large outflow of artesian water, and although the successive Governments and the Department were apathetic in the matter for too many years previous to the advent of the successful Government bore at Barcaldine, the Department is to be congratulated on the ulterior great success of its operations, which operations since they were started, it need scarcely be said, have been prosecuted with great vigour and intelligence. Twenty bores, averaging about 1600 feet each in depth, have been sunk, and out of this number, although the Department has done the pioneering, in the deeper bores, only one of those completed to the contract depth has been a failure.

One station (Uanda) has twenty-nine bores, water being obtainable at depths ranging from 194 feet to 753 feet. Amongst the deepest bores in the Colony are Malvern Hills bore, 3948 feet, and not completed; Winton, 4004 feet; Wellshot, No. 1, 3500 feet; Dagworth, 3335 feet; Chatsworth, 3130 feet, and still in progress; and Bowen Downs, No. 4, 3109 feet. All these, with the exception of Winton, are private bores, the cost of which, although necessarily proportionately high, has been more than compensated for by the great and inestimable value of the water obtained.

The most completely successful bore to date appears to be Boatman No. 1. Reference to the list of bores will show the depth of this bore to be 1500 feet, and the outflow 4,200,000 gallons per diem. An analysis of the water* states that it is perfectly good water for all purposes. One of the latest additions to the artesian well system of Western Queenslsnd is a supply of over three quarters of a million gallons daily of pure water struck on the 2nd May, at a depth of 1015 feet The bore is on Jacondal Grazing Farm near Barcaldine. The contract was made on March 23rd. On April 2nd, the plant was carted across country from Home Creek run 33 miles, and unloaded on the new site. The derrick was built and machinery put in position in ten days, and boring commenced on April 12th. From this date to the 18th April, the bore was put down to 530 feet, where a supply of water of 1200 gallons per diem flowed over the top. A long delay was occasioned by the

---

* See Section "Analysis of Bores Water."

caving-in of the sides of the bore, which protracted operations until the bore was cased. In 48 hours afterwards a depth of 700 feet was reached and the remaining 300 feet were bored in the short time of four days and a half, the full supply of 800,000 gallons per diem being obtained at 1015 feet, or in ten and a half days of actual boring from start to finish.

Section XII.

PIONEER BORES OF AUSTRALIA.

UNDER the head of "Definition of Artesian Water" will be found a further definition of it by the Queensland Government Hydraulic Engineer, Mr. Henderson. I quote fully from Mr. Henderson in confirmation of a similar definition which I have always held myself. Experience has shown that wells may be "flowing" in one part of a run, and "non-flowing" in another part, although in the latter the water may rise to within a few feet of the surface and even in some seasons overflow it. Undulation or difference in the level of the surface may produce this result, although the water in both wells be derived from the same true artesian source. And this raises a question of some moment, *i.e.*, to whom the priority or credit is due as discoverers of artesian water in these colonies. It would seem to be manifestly unjust to award the palm to any particular bore if it be taken into consideration that the same degree of skill and energy may have been evinced in making a "non-flowing" as a "flowing" well, when the water struck in either case is from the same artesian source.

The first subterranean water—which was sub-artesian—found in Australia of which I can find record was obtained in 1878, at Mogg's Plains, St. Arnaud, by machine boring by the Victorian Government. The first sub-artesian water obtained by the same means in Queensland, was struck in 1882, in a Government bore at the "14 Mile," near Cunnamulla, and the first artesian water was obtained by the Railway Department at Back Creek, twenty miles from Barcaldine, on the Central Railway. and in the Government bore at Barcaldine, in 1887. The first (Government) artesian bore, that at Blackall, was started previous to the one at Barcaldine, in January, 1886, but it was not completed until some time after the Barcaldine bore had proved successful.

According to the Report—1891—of the Water Conservation Department of New South Wales, Mr. J. W. Boultbee, in charge—the first artesian bore made in that colony, was at Kallara Station, in 1879. It was made to a depth of 140 feet only, when artesian water was struck that rose 26 feet above the surface, and it has been running without intermission ever since. This appears to be the first artesian bore of which there is record in Australia. In 1880, two years after my first communication to the Press (the Melbourne *Leader*) on "Artesian Wells for Australia," a bore was put down at Sale, in Gippsland, Victoria, in which artesian water was obtained at a depth of 234 feet, the flow rose 16 feet above the surface, and the supply was 36,000 gallons per diem. The water was somewhat saline, but it is a fact worth recording, that after exposure and its attendant aeration, it became fresh.

The Machines first used by the Victorian, New South Wales, and Queensland Governments, before the deeper artesian water was obtained, were principally the "Tiffin," and the "Wright & Edwards," boring machines, the capacities of which were limited to a few hundred feet of boring only, and no deep artesian water was, so far as I can ascertain, obtained excepting in the bores alluded to at Kallara, New South Wales, and Sale, Victoria, which were made by private parties.

The subsequent experience gained by Queensland, after the advent of the deep artesian machinery, shows that a great deal of most valuable artesian water had existed at moderate depths, but had been missed by the early borers, as follows (*vide* also Appendix A) :—

	Depth. Feet.	Gallons.
Richmond Downs, No. 2	480	500,000 per diem.
Currawinga	264	100,000 ,,
Dynevor Downs	⎧ A number of bores 86 feet to	
Bingera	⎨ 210 feet, with a flow of 40,000	
Manfred Downs	⎩ to 70,000 gallons per diem.	
Uanda No. 1	335	500,000 per diem.
,, ,, 2	362	150,000 ,,
,, ,, 3	248	80,000 ,,
,, ,, 5	280	30,000 ,,
,, ,, 7	226	100,000 ,,
,, ,, 8	509	200,000 ,,
,, ,, 13	440	250,000 ,,

In view of the greater success of the deeper bores it is scarcely necessary to point out how much more politic it would have been had the Governments entered upon deep boring at once without hesitation or delay, and have thus followed the advice very freely tendered at the time, of artesian well-boring Engineers (who had in the meantime to put up with limited apparatus) and have thus avoided the incalculable losses in stock from which the country had previously suffered during every succeeding drought.

SECTION XIII.

DEFINITION OF ARTESIAN WATER.

AN artesian well is a shaft bored through impermeable strata until a water-bearing stratum is reached, when the water is forced upwards by means of the hydrostatic pressure due to the higher level at which the main or supply water was received. The action of an artesian well depends upon very simple principles. The water accumulates and is conserved in porous rocks and ground lying between two layers of impermeable strata forming a basin, which may be of very considerable extent. These porous rocks crop out at the surface, and form thereby the means of intercepting, on the higher levels or outskirts of the basin, the rain and flood waters, which sink into them. Becoming surcharged with this water any boring which is made from the surface to the lower impermeable rock will, at greater or lesser depths, intercept this water, which from the hydrostatic pressure of that part of the accumulated water above the surface of the ground at the site of the boring will rise above the surface to the highest level at which the accumulated water stands. A simple illustration is given by taking a tube of the form of the letter U. If water be put into this tube it will stand at the same level in both arms, and if a third tube be connected with it between the two arms, the water will rise in it and sink in the other two until equilibrium is restored, and if the third tube be shorter than the height at which the water so stands an outflow will take place from it either continuously, if the supply be maintained, or until equilibrium is restored. The laws upon which the action of an artesian well is regulated are therefore:— (1) The orifice of the bore must be below the outcrop of the water-bearing stratum. (2) The water must be contained between two impervious strata. (3) The strata must take the form of a basin, or the outflow must be so far impeded as to keep the water at an elevation higher than the orifice of the bore. These are the simple principles upon which the action of artesian wells depend.

In the Queensland Government report of the Water Supply Department for 1893 Mr. J. B. Henderson, Hydraulic Engineer, in treating upon this subject has, I think, very properly further defined the word, "artesian." He says:—"Strictly speaking, the term 'artesian well' means one that is bored vertically through superincumbent impermeable strata to a lower stratum

Definition of Artesian Water.

highly charged with water under such conditions of pressure, owing to the superior elevation the outcrop has above the well, as will cause the fluid to ascend through the bore and to flow on the surface, or to rise in the well to a height from which it can be conveniently pumped; hence, by this definition, whether a well is a 'flowing one' or not it may be an artesian one. In this connection it has been suggested that a 'flowing well' should be distinguished simply by the word 'artesian,' and that non-flowing wells tapping artesian water should be marked by the term, 'sub-artesian.' I favour these terms, and generally use them."

I think the above further definition should be adopted in future, and it is doubtless based upon the evidence given in the report of such Queensland bores as the following :—

	Depth in feet.	Yield per diem in gals.	Remarks.
Manfred Downs, No. 1 Bore ...	177	22,000	Overflowing.
„ „ 5 „ ...	200	20,000	Not overflowing, supply pumped.
„ „ 6 „ ...	98	16,000	Overflowing.
Thurulgoona 2 „ ...	1440	30,000	Overflowing slightly, supply pumped.
„ 4 „ ...	718	36,000	„ „
Uanda 6 „ ...	296	40,000	Water rises to within 35ft. of surface.

The waters of these bores are undoubtedly derived from true artesian supplies.

Mr. Henderson further remarks in the report :—"Generally stated, artesian water is from a different source than that obtained from 'surface wells,' which derive their supplies from shallower strata, such as sands, gravels and other porous beds that are surcharged with water, the permanence of which is immediately dependent on rainfall; hence the supply of water from surface wells oscillates with the seasons. It often becomes much reduced or fails altogether in dry weather, or when it is largely drawn upon."

My practice has given the same results as regards "flowing" and "non-flowing" wells. In cases where the bores were not actually "overflowing" ones the water rose to within from 15 to 40 feet of the surface.

Undulation, or difference in the level, of the surface will give a flowing well in one part of a station and a non-flowing well in another part, the water in both of which is derived from the same artesian supply; and this raises a curious question—that of the priority of the discovery of artesian water in these Colonies. It would seem to be manifestly unjust to give the palm to a particular bore in which the water barely overflowed —as in the case of the Thurulgoona bore, just mentioned, the water in which has to be pumped—as against another previously made in the immediate vicinity in which the water rose to within a few feet of the surface, when both these waters being derived, as they are, from an artesian source are truly artesian and have involved an equal degree of skill and energy to put them down. Cases have occurred in the United States where "flowing" wells of little pressure have become non-flowing ones, and in previously non-flowing wells in the same district the water has afterwards overflown the surface.

Section XIV.

THE HOTTEST REGIONS ON EARTH.

In the Eastern Hemisphere the hottest spot is on the borders of the Persian Gulf, on the south-western coast of Persia. The thermometer during July and August never falls below 100 deg. Fah. during the night, while the temperature during the day rises to 128 deg. or 130 deg. Little or no rain falls, and yet, in spite of this terrific heat and other drawbacks, a comparatively numerous population contrive to live there, obtaining their water supply by divers from the copious springs of fresh water which burst forth from the bottom of the sea, these sources of supply being originally derived from the rain water, which sink into the porous strata inland and discharges itself at the lowest level attainable—the bed of the ocean.

In the Western Hemisphere the hottest region is a valley in California (known as the Death Valley) situate to the east of the Sierra Nevadas and running between two mountain ranges, the Tuneral (6000 feet) and the Amargosa (10,000 feet), which has a higher temperature than the region on the Persian coast. In four months out of five during which readings of the thermometer were taken, the mean temperature rose above 90 deg., while in July and August it exceeded 100 deg. The mean temperature for the twenty-four hours on the 18th July, 1891, was just over 108 deg. This valley is uninhabited, and derived its significant name from the circumstance that an active party of Californian emigrants, who had strayed from the regular overland route, perished there in 1850 from heat and thirst.

The hottest region in Africa is in the Nubian Desert, where food may be cooked by being buried in the sand. The Arabs say of it, "The soil is like fire and the wind like a flame." The hottest portions of the British Empire are India and Australia.

The Sierra Nevada Valley in America must, from its physical conformation, be doomed to oblivion so far as improved water supply is concerned; but not so the African Nubian Desert, which is destined, doubtless, in time, under the influence of British colonisation and the borer's drill, to vie with the Great Desert of Sahara, which, as mentioned, has undergone a wonderful transformation since the advent of its artesian water supplies.

Section XV.

ARTESIAN BORES AS AFFECTING CLIMATE.

The late Dr. Bancroft propounded the question as to whether the action of the water from numerous bores in the interior of the country would induce moisture sufficient to coalesce with the rainy atmosphere of the coast and thus determine rainfall in that interior. This question has often occurred to me in my practice, which has been in the driest parts of the interior of this country.

In all lake districts, with their accumulation of inland surface water, the rainfall is greater than in the adjacent waterless country. This I found to be the case during a visit I made a few years ago to the chain of reservoirs which supply one of the largest cities in Great Britain with water. With a heavy periodical rainfall it was found that during spells of dry weather in the adjacent country rain fell invariably more or less over the line of the reservoirs. I remarked the same effect in many sections of the United States, particularly in Southern Illinois and Iowa—representative farming States—where irrigation from bored wells on the closely settled farms induced a larger and more regular, although light, rainfall than that which took place on the unsettled prairie lands of the country.

Section XVI.

GEOLOGICAL CONDITIONS AFFECTING THE FLOW OF ARTESIAN WATER.

The accompanying diagrams will show the various forms in which artesian water exists in the geological strata to which it belongs.

Certain portions of the water of every creek or stream are always "running uphill" though its average mass is moving down. The bottom layer flows up and down according to the inequalities of the bed, while the top layer declines more uniformly with the surface slope. In the artesian stream we only see the rising column issuing from the earth and the creek that flows away. The more potent descending volume that forces the flow is concealed; no portion would rise if it were not forced up by a superior portion pressing down.

Fig. 1

Figure 1 of the diagrams is a longitudinal section of a stream, illustrating in part its upward current, subordinate to the general downward flowage. To form an idea of the common class of flowing wells imagine a pervious stratum through which water can readily pass. Below this let there be a watertight bed, and let a similar one lie upon it, so that it is securely embraced between impervious layers. Suppose the edges of these layers to come to the surface in some elevated region (save that they may be covered with soil and loose surface material), while in the opposite direction they pitch down to considerable depths, and either come up to the surface again at some distance (figure 2), or else terminate in such a way (figure 3), or take on such a nature (figure 4) that water cannot escape in that direction. Now let rainfall and surface water penetrate the elevated edge of the porous bed and fill it to the brim. That such beds are so filled is evident by ordinary wells, which commonly finds a constant supply at no great depth. Now it is manifest that if such a watered bed be tapped by boring at some point lower than its outcrop, the water will rise and flow to the surface, because of the higher head on the upper edge of the bed. If the surface water continually supplies the upper edge as fast as the water is drawn off below the flow will be constant.

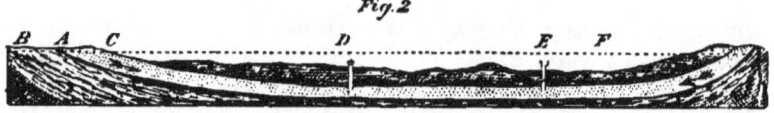

Fig. 2

Figure 2 is an ideal section illustrating the chief requisite conditions of artesian wells. A, a porous stratum; B and C, impervious beds below and above A, acting as confining strata; F, the height of the water level in the porous bed, A, or, in other words, the height of the reservoir or fountain head; D and E, flowing wells springing from the porous water-filled bed A.

Fig. 3

Figure 3 is a section illustrating the thinning out of a porous water-bearing bed, A, enclosed between impervious beds, B and C, thus furnished the necessary conditions for an artesian flow, D.

Fig. 4

Figure 4 is a section illustrating the transition of a porous water-bearing bed, A, into a close-textured, impervious one. Being enclosed between the impervious beds, B and C, it furnishes the conditions for an artesian fountain, D.

Fig. 5

Figure 5 is a section illustrating the usual order in which the strata of a basin come to the surface. A and B, porous beds; D and E, impervious beds; C, a half impervious bed; the dotted lines are the water levels of A and B respectively.

Geological Conditions affecting Flow of Artesian Water. 77

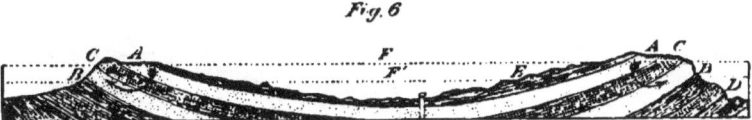

Fig. 6

Figure 6 is a section illustrating the possible effect of erosion upon strata originally like those in figure 5. A and B, porous beds; D and E, impervious beds; C, a half pervious bed; F and F, the water levels of A and B respectively. If the stratum C is not practically a confining layer, the water from it will pass through it and escape at the edge of B, so that a flow cannot be obtained at a higher level than it, but may be had below the line F.

Fig. 7

Figure 7 is a section illustrating the failure of an artesian well because of defects in a confining bed below. A and B, porous beds; D and I, impervious beds; C, a defective confining bed; E, the water level of the stratum, B; G and H, wells that do not flow. The bed A might give a flow at G and H but for the defect in C, which permits the water to descend into B and escape through its outcrop, which lies below the surface of G and H.

Fig. 8

Figure 8 is intended to illustrate the aid afforded by a high-water surface between the fountain head and the well. A, a porous bed; B, a confining bed below; C, a confining bed above. The dark line immediately below the surface represents the underground water surface. Its pressure downward is represented by the arrow m. The pressure upward, due to the elevation of the fountain-head, is represented by the arrow n. The line F represents the level of the fountain-head. There can be no leakage upward through the bed C except near the well D. There may be some penetration from the bed C into A, which would aid the flow.

Fig. 9

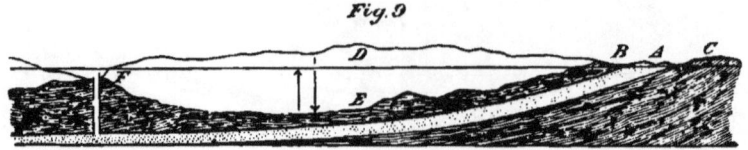

Figure 9 is a double section illustrating the effects of high and low water service in the cover area. In this A represents the porous stratum enclosed between impervious beds, B, C. The source of supply is at A, and proposed well at F. Let E be supposed to represent surface of ground in one of two supposed cases, and D the surface in the other. The arrow springing from surface E represents upward tendency of water in porous bed, owing to pressure from fountain-head, while the arrow depending from line D represents downward pressure of ground water, whose surface is represented at D, and is, it will be observed, more than equivalent to the upward tendency due to pressure from the fountain-head. A flow at F could be very safely predicted if the surface were as represented by D, while it might be doubtful whether one could be secured if the surface were as represented by E.

Fig. 10

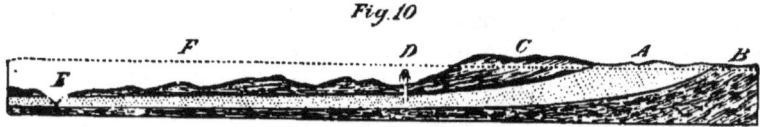

Figure 10 illustrates the possibility of a flow from a bed even when exposed at a lower level. A, a sandstone bed, thick and coarse at the right, its shore edge, and thinner and finer at the left; B and C confining impervious beds; F, the water level in A; D, a well which may flow notwithstanding the lower exposure at E.

Fig. 11

Figure 11 shows the dependence of the collecting area on the thickness and slope of the porous beds. In the left hand figure the porous bed (A) is thin, and coming to the surface at a high angle gives but a small section. In the right hand figure the bed (A) is thick, and, coming to the surface at a low angle, its bevelled edge is broad.

Geological Conditions affecting Flow of Artesian Water. 79

Figure 12 illustrates a common effect of erosion upon the surface area of the porous stratum, and the contour of the resulting basin. The dotted lines show the original contours.

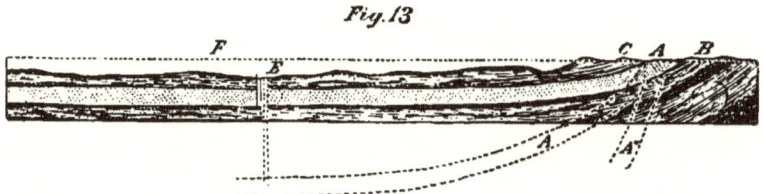

Figure 13 is a section illustrating the advantages of the low inclination. A, porous bed; B and C, impervious beds; A¹ and A¹¹, indicate porous beds of higher dip.

As pointed out under the head of "Permanence of Supplies," the recently-published Transactions of the United States Geological Survey show that the presence of artesian water depends upon physical laws, which may be present in any geological formation, providing the required conditions to impound it and force it to the surface are present, and that it is not necessarily confined to any particular formation, either sandstone—similar to the Clydesdale braystone—or any other.

Sections of Two Representative Australian Artesian Bores, with particulars of Strata.

CHARLEVILLE BORE, QUEENSLAND.

		Feet.	Depth in ft.
(A)	Sand-clays and gravels of the Upper Cretaceous formation	270	
(B)	From 175 feet to 195 feet ; excellent fresh water in gravel, did not flow		
(C)	Clays, sandstone, and light-coloured shales of the Lower Cretaceous formation, interstratified with thin beds of gravel	310	875
(D)	At 310 feet a further supply of good water.		
(E)	Well-defined blue shales of the Lower Cretaceous formation ; artesian water struck at 1,371 feet ; flow, 3,000,000 gallons per diem ; water clear, colourless, soft, and potable ; temperature, 106 deg. Fahr.		1371

MUCKADILLA BORE, QUEENSLAND.

	Feet	Depth in ft.
(A) Yellow clay	50	—
(B) Black shale	370	—
(C) Sandstone	24	—
(D) Black shale	234	678
(E) Gray pipeclay	116	794
(F) Gray sandstone	195	989
(G) Sand drift, white, and lignite, tapped water	21	1010
(H) Shale, brown and gray	61	1071
(I) Sandstone conglomerate, gray ...	26	1097
(J) Shale, brown and gray	54	1151
(K) Sand drift, white and brown, water increased	127	1278
(L) Shale, brown	15	1293
(M) Sand drift, gray	32	1325
(N) Shale, gray, silicious...	123	1448
(O) Sandstone, brownish gray	85	1583
(P) Sand drift, white and gray	148	1681
(Q) Shale, blackish	150	1831
(R) Sandstone, gray	29	1860
(S) Shale, slate-coloured to brownish gray	98	1958
(T) Shale, black, with streaks of coal ...	277	2235
(U) Sandstone and shale, soft	265	2500
(V) Sandstone, white ; water flowing 10,000 gallons per diem	618	3118
(W) Sandstone, micaceous, overflow of water, 1290 gals. per day at 3170ft.	52	3190
(X) Sandstone, white and soft ; water overflowing 23,000 gals. per diem ; temperature, 124 deg. Fahr. ...	92	3262

SECTION XVII.

PERMANENCE OF SUPPLIES.

THE question of the permanence of the artesian supplies is, it will be readily understood, of vital importance. The following treatise bearing upon the subject was published for me in the Melbourne *Leader* in 1878 and in the *Brisbane Courier* in May, 1883, three years before the first Queensland artesian bore (Government), that at Blackall, was commenced, and it will be found equally applicable at the present time :—

" The geological formation of these artesian basins varies in different localities, but, broadly speaking the lower impermeable or Primary bed-rock consists either of Silurian, Metamorphic, or granite rocks. The newer formations which fill in the basin or trough are various in character. Cretaceous or Upper Mesozoic rocks overlie the Palæozoic rocks or Miocene or Middle Tertiary formations. The bed rocks are practically impervious to the escape of water in large quantities, but in all the newer formations there exist cavernous spaces through which water can readily percolate, and it is in these newer formations that search for artesian water will prove successful.

" The Australian continent may be described as a huge basin, the edge of which consists of an elevated coast range. The interior of this is almost entirely filled in with the newer formations; the isolated ranges with their peaks protruding through them consisting of the older or bed-rock formations alone. During ages of time the action of flowing water from the coast range, and from the secondary ranges of the Australian interior, has been producing constant changes in removing the denuded surface, making valley-like depressions on the surface, through which the excavated material has been carried and deposited at lower levels. By these means the great central basin has been formed, and of this there is ample evidence visible in the interior, in the mud plains, in the river channels, and in the frequent remains met with of what geologists call Desert Sandstone. Denudations of still older formation have been also going on, which have been planed off down to the vast granite masses that everywhere form the foundation of the Australian Continent. In this great embracing basin are interior and local ones lying, as before stated, between the outcrop of the impermeable rocks in which the true

G

artesian water will be obtained. Allowing for evaporation and the quantity of water passing down the surface rivers of this Colony, that passing away underground after the basins have become surcharged must be enormous. The great tropical rainfall of the north, the rapidly-absorbing nature of the soil over vast tracts, the short rivers or creeks which eventually lose this water by soakage, all tend to show that a very large percentage of the water passes away underground to the ocean.

"The strata from which good results will also doubtless be obtained are those alluvial beds of sand and gravel which are in the leading valleys and extend to very considerable depths, covering the sites of ancient watercourses. It is also probable that the great plains of the interior may contain Tertiary beds of great thickness in which an abundant supply of water could be obtained. Wells have been made through clay and marl, when a plentiful supply of water has been obtained. In some localities a limestone formation exists overlying sandstone. This formation would be very likely to afford large supplies of water by means of bores, as the rain upon the surface of these generally flat and sandy districts percolates through porous beds of sandy clays, gravel and sand down to the limestone beneath."

In January, 1895, before the publication of the Government Geologist (Mr. Jack's) paper (see section "Queensland"), *The Brisbane Courier* published the following, written by me, "On the Permanence of Supplies and proposed legislation to regulate the outflow":—

"Certain points in a previous Government Report—that of 1892—were discussed in July, 1893, in your columns by Messrs. Burkitt (Maxwelton Station), Wienholt (Warenda Station), myself and 'Spinifex' in the 'Pastoralists' Review,' and by others. These points were mainly—(1) The extent of the areas of the outcrop, or intake, of the artesian water-bearing rocks; and (2) The possibility of exhaustion of the supplies and Government control of the same. In previous reports the Hydraulic Engineer gave it as his opinion that the outcrop could not be more than a few chains wide, giving 200 square miles as the probable area (see the Government Geologist's views, section 'Queensland'), and as he thought there was a doubt as to the permanence of supplies, he suggested legislation to regulate the free use of the water from the bores. The last Report, 1894, says 'little can be added as regards the probable area of the outcrop of the water-bearing rocks' (the Cretaceous). The public is, therefore, left in doubt upon this point. An experience

Permanency of Supplies. 83

at first hand gained in the United States during a residence of some six years, from which country our inspirations and experience upon the subject of artesian water are really and unavoidably drawn, showed me clearly that it is not in one particular strata of rock alone that artesian water may be obtained. This has been amply verified by American experience during the last thirty years. The recently published 'Transactions of the Geological Survey of the United States' make no reference to the Cretaceous system, or any other particular system, as being water-bearing more than another. 'As a matter of fact,' says the 'Transactions,' 'the presence of artesian water depends upon physical laws which may be present in any geological formation, providing the required conditions to impound and force it to the surface are present.'

' A great number of the artesian wells of the United States are not in what geologists would call the Cretaceous system, but notably in the Carboniferous formation, the sandstones, &c. This being the case, we have, in all probability, in Australia an immensely larger area of outcrop of water-bearing rocks than that limited to the Cretaceous formation alone, as implied in the Government report. One of the largest items in the expenditure upon deep coal mining in Great Britain is for pumping and subduing the fresh water lying in the deep sandstone rocks of the Carboniferous system met with in sinking the shafts. At Monk Wearmouth Colliery, Sunderland, England, the shaft was sunk to a depth of 1,000 ft., when a spring was tapped, giving 3,000 gallons per minute, equal to over 4,000,000 gallons per diem of good water.

"As to the permanence of supplies and State control of the wells, we must call for information and experience from which to profit—as we have already done in regard to deep-boring appliances—upon the American States. America has done the pioneering on a very extensive scale, and over a long course of years, in the art of obtaining underground water. The American Reports just to hand confirm in a most satisfactory manner the conclusions I was able to arrive at during my residence in the States, and show conclusively that indisputably great supplies are maintained, which provide for the outflow from the artesian bores, those bores being in free and uncontrolled operation. As a recent traveller, of great Australian experience, remarks ' many thousands of wells will have to be drilled in these colonies before we can obtain an amount of water anything comparable to the quantity that the Americans have been drawing from their wells for many years past.' The total number of artesian wells in the States, as given by official statistics recently published, is nearly 17,000. These wells at an average depth of 1,000 ft. per bored well gives about 3,240 miles of boring.

"Legislation has once or twice originated in Queensland with a view to control the outflow from bores on private holdings, and a bill entitled, 'For the Regulation of Artesian Wells, and for Preventing the Waste of Artesian Water derived therefrom,' has been introduced to the Assembly of New South Wales. It places most stringent restrictions upon the artesian water supply of that colony; heavy penalties being proposed on persons who may—'in the opinion of an officer,' to be appointed under the act—improperly use the water. It is advanced by the promoters and adherents for this State control that it is absolutely necessary in order to prevent the exhaustion of the supplies, and the contention is supported by the assertion that the United States Government has been induced to pass restrictive legislation in this direction. As regards Australia, the entire extent of the supplies is not proved one way or the other. We have no evidence that our supplies are failing, but we are, notwithstanding, advised on uncertain grounds to legislate.

"As to the practice in America, it can be authoritatively stated that neither in the States of California or Dakota, which are pre-eminent for their large supplies of artesian water, nor the other States of the Union, is there legislation, either locally or by the Government, at Washington for the control of either public or private bores. South Dakota has passed laws for the promotion of irrigation, but not for limiting the flow of water from artesian wells. Chapter 109 'Irrigation,' of these laws regulates the boring of wells under State supervision. Bonds may be issued by the township, and taxes levied for their payment; but the State is not directly interested. It is believed that the development of the country is better left in the hands of private citizenship, and the law is intended to help those who are not able to help themselves. The greater number of wells in this State have been made by individuals, and they have entire control of them. In Utah and Colorado periodical attempts have been made to procure State legislation against waste of water, and to put a stop to the drilling of an excessive number of wells, and in cases where local laws have been passed they have not been enforced; but it must be borne in mind that the artesian basins in these States are small, and the Government authorities look upon the partial failure of any small artesian basins as compensated for by the undeniable success during a long course of years, of the deep wells of the semi-arid regions

Permanency of Supplies. 85

of the other States. The Government have instituted no control whatever of the water found by individuals in the States, the bulk of which, as artesian water, has been running freely for many years, such action being at the same time incompatible with American political institutions.

"We in Queensland have drawn upon America and wisely so, too, from the very earliest date (1881) of the movement in this colony for a subterranean water-supply, for information and for the necessary machinery and the skill to work it. Surely we cannot be far wrong in drawing further upon American experience, so freely and courteously offered to us, in the subjects under consideration."

I have shown in the above communication the experiences of the Americans of the permanence of artesian supplies. From my own extended observations in Australia—particularly in Queensland—so fully confirmed as they are by the Queensland Government Geologist in his recent further explorations, I am convinced that there is a very large quantity of water lying deep down in the crust of this portion of our earth, which will admit of a very large draught upon it, and so long as the rain falls it will be, in my opinion, amply replenished, providing for all the outflow it will be called upon to make for very many years to come.

Section XVIII.

IRRIGATION FROM ARTESIAN BORES.

IRRIGATION from artesian bores has long been an established practice in other countries, especially in India, Algeria and America, in which it has been carried out on a large scale.*

In India the value of local irrigation is shown by the Indian Revenue Reports. The land irrigated by wells in the Madras Presidency is about 2,000,000 acres in extent, yields a revenue of £1,500,000 sterling, £2,000,000 sterling being the revenue from 25,000,000 acres of non-irrigated farm land. In one case the land revenue is twenty-seven shillings per acre, and eight 1.6 shillings in the other. The intrinsic or market value of crops from irrigated lands average thirty-four shillings per acre as against ten shillings in the non-irrigated. The latter depends entirely upon the local rainfall.

It is a fair estimate that there are in the world to-day at least 200,000,000 persons depending solely for their food upon areas irrigated by water drawn in the most primitive manner from underground sources in the form of wells, springs, or drainage conduits existing in Central Asia and Persia. An examination of the records, habits, and customs of the communities so supplied will show an elaborate system of care and maintenance. The countries in which this system has been most widely obtained have in past centuries been more highly civilised and have borne a large share in the ancient history of the world. The constant and unchanging tendency of climatic aridity has conquered the ameliorations produced by human industry, filled up the conduits, choked the wells, destroyed the surface works, and from want of the cultivation previously existing has produced and intensified the vast desolation which now exists over such large areas. The re-establishment of the irrigating works and the conservation of the water supplies required therefor will quite certainly restore enormous portions of these great areas to the use of man. If such things can be done under a tropical sun they will certainly be achieved under the milder general climate of Australia.

M. Rolland, in the *Révue Scientifique*, of June, 1886, says respecting the Algerian irrigation from wells:—"The common wells of the Algerian Sahara are quite numerous, especially in the centre of the region, but the prosperous sections are those regions which are supplied by artesian waters. It is not necessary

---

* A partial description of this mode of irrigation in these countries is given severally in their respective sections.

that these waters should gush or flow above ground, for in some parts of the Sahara and the province of Constantine there is found a sheet of water reached from the surface at a depth of from ten to twenty feet beneath. When tapped this supply rises rapidly. The area in which it is found is close to, and sometimes below, sea-level. The very gardens and vineyards that are cultivated by this supply are lodged in little cavities dug out of the soil between the sand dunes that cover the larger portion of the Souf region. By this means the palm roots itself in the watery bed found below, thus obtaining a constant, though moderate, humidity. The earth dug out forms a barrier around each of these little gardens, against which the sand accumulates, protecting the palms from being overwhelmed, and making, at the same time, a curious feature of the landscape, as their green plumes seem often level with the ordinary surface. Under these conditions the palm comes to perfect maturity. Many thousand acres are reclaimed in this way. In the artesian regions proper the water gushes forth with a small, but steady and permanent flow. The manner of using it in all sections is to distribute it in proper channels, it being constant at all seasons of the year. Within a few years past this system of artesian waters and distribution has penetrated eastward into the Sahara of Tunis, and considerable areas are under cultivation by this method. A remarkable fact in the development of this supply is the enormous increase which has resulted from the systematic boring of wells and the distribution of waters. So large is the supply in some parts that *colmolage*—that is, draining with irrigation—is necessarily practised to ensure the security of the crops and the health of the inhabitants. The Sahara regions of Southern Algeria and Tunis possess immense basins of artesian water, no doubt with high pressure, and flowing therefore with considerable force. In L'Oned Rir the united discharge of these artesian wells amounts to 34,346,000,000 gallons a year, or 213,714 gallons per minute. At the time of the French conquest of this region the condition of these oases was most lamentable. Cultivation was fading away from want of water; many of these oases had disappeared entirely under the sand clouds which are constantly blown across the surface of the Sahara, but now they are well cultivated."

"The conservation of water by the Algerian system of well-boring and storage is strikingly illustrated by the following table of annual use and waste in acre-feet (325,829 gallons of water per acre, or 1 cubic foot upon the surface of an acre of land) at the village of Saint-Denis du Sig:—

Date.	Waters Utilized. Acre-feet.	Not Utilized. Acre-feet.
1863	9,351	2,397
1864	8,893	5,955
1865	16,235	33,410
1866	10,485	—
1867	4,511	—
1868	13,594	7,489
1869	10,757	6,106
1870	10,250	3,644
1871	11,578	10,250
1872	16,485	18,385
Total	112,139	87,636 "

Irrigation by means of bored wells has nowhere received such practical and favourable economic results as in the United States of America. In the State of California, with a climate similar to that of Australia, operations have been carried out on the largest scale. As stated under Section "American bores," "There are upwards of 3,000 bored-wells in that State irrigating from one-quarter acre to 1,000 acres each. They exist from San Diego to Shasta County, and while their flow is not one-tenth utilized, probably 60,000 acres are at present served. Their possible development in arid regions is almost illimitable."

At present the developed basins of that State are :—

Those of the high altitude areas, such as that of Sierra County where 200 wells are already flowing ;

Those of the great Central Valley or inter-mountain plain of Sacramento and San Joaquin ;

Those of the San Francisco Bay region extending from the southern end of the bay in a triangular shape down the coast to an apex at San José ;

Those of the coastal region south from San José, with its table-lands, valleys, foot-hills ;

Those of the extreme southern counties of Los Angeles, Orange, San Bernardino, and San Diego, including also the desert and plateau lands of the Colorado Basin, the Mogava Desert, and the Antelope Valley.

Thousands of bored wells are in existence, the water of which rises nearly to the surface, and myriads of windmills can be seen all over the landscape engaged in pumping this water for the use of farm-yard and field. A very considerable area of which no direct record has been made, now in use for garden and orchard purposes, is maintained in fertility by water drawn from this

AMERICAN IRRIGATION FROM ARTESIAN WELLS.

A Seven-inch Bore Artesian Well, 450 feet deep, Paige and Morton's Ranch, Tulare, California.

source. In the southern coast counties these wells are very numerous. The drainage waters of the coast range are also tapped in the low rolling foot-hills by the enlargement of the springs or the driving of small tunnels therein. The California fruit-grower is just beginning to apprehend the importance of this supply. Two of the most important artesian belts known to the world are found in the San Joaquin Valley and in the counties of the State south therefrom. Tulare and Kern Counties offer interesting facts. The vast bulk of Mount Whitney towers over the Sierras to the east thereof, forming the central point of that enormous range,* and sending down surface strata and 80 to 100 inches of precipitation which annually falls upon its summit. The swollen floods of the King, Kern and smaller streams in their turn supply the great canal systems of Kern and Tulare. Tulare Lake is fed from this source. The whole of that section of California's great valley will doubtless be found, when proper investigation is made to be underlaid with the same supply; and in a belt of country lying to the north and south of Tulare Lake, some 40 miles in length and 50 in width, over 100 flowing wells have already been bored. South of Tulare, and in Kern County, this belt or a similar one runs north and south for a distance of 50 miles and from 10 to 15 wide. In the Tulare basin the average depth of 40 of the principal wells is 469 feet, and of the remainder about 300 feet. At that depth they are considered to possess the best and strongest flow. The lowest flow for 24 hours is 200,000 gallons, and the highest 3,000,000, while the average is about 1,600,000. In the centre of this tract, which contains the strongest wells, an area of 18 by 14 miles, the total flow is estimated at over 48,000,000 gallons per 24 hours. It is stated that 2400 gallons flowing continuously for the season of 100 days will irrigate one acre of land. If the water from the wells producing this flow was properly stored, it would, allowing the loss of one-third by evaporation and percolation (which for well water is a large estimate), irrigate at least 42,000 acres of land. There are now from 10,000 to 12,000 acres served by the Tulare County wells. Of 40-acre farms 4000 could easily be supplied from this tract without pumping or other mechanical means of lifting. A flow of six inches over an 8-inch casing, which is a common feature of these wells, will

---

* These mountains may be higher than the Australian ranges and the rainfall on them may be greater, but it will be remarked that the Queensland bores especially show a much greater maximum and average discharge than any of the American bores, so that practically the existence of high mountain ranges and greater rainfall do not appear to influence a greater outflow than lower ranges and less rainfall.

give five acre-feet for every 24 hours, and serve 350 acres of land. The Kern County belt is even more remarkable. The engineers who have been employed by the owners of the great canal system estimate that the artesian underflow will cover the entire valley region, a tract of country as large as the States of Delaware and Rhode Island combined, and capable of sustaining by agriculture a much greater population than those two States. The average Kern County well will serve, it is estimated, one section of 640 acres of land. If its continuous flow is properly stored, its service can be made as high as 3500 acres. If these estimates are correct, the Kern County artesian belt will become in its economic quality the second important in the world. So strong is the flow when the artesian supply is struck, that the farmers find it difficult to control it.

In the upper part of Kern County, just south of Tulare Lake, there are about 40 of these great wells within a radius of 10 miles, none of which yields less than 1,000,000 gallons per day. The 12 largest, whose flow ranges from 1,200,000 to 2,500,000 gallons, give a total flow each 24 hours of 23,600,000 gallons, which will serve at least 20,000 acres. The average depth of the upper or northern wells is about 350 feet, ranging from 250 to 460 feet. Elsewhere the range is from 325 to 1000 feet in depth, the cost running from £100 to £600, and averaging about £280 each. The use of these wells has been confined to the cultivation of alfalfa and other cattle feed. There are probably in the same region 500 or more bored wells, the water of which, being derived from stratum near the surface, does not rise thereto, but is lifted above it by mechanical means. The deeper wells feed irrigation ditches of many miles in length and of considerable depth and width, which have the appearance of small rivers. Kern County has been controlled chiefly by large landowners, whose interest now compels the adoption of the colony or small holdings system. With the increase of this form of settlement there will come a rapid exploitation of the under-flow water supply, and more especially that the surface canal system, which has been created at the expense of several millions of dollars, will probably for some time remain in private hands to be used as a means of obtaining a large rental income. Prosperity on the part of the settlers will compel them to seek independence from this control, and as a result the great artesian underflow will be rapidly utilised.

An examination of the various American Government reports will show that, independent of extensive irrigation from artesian wells in the older or Eastern States and in the more settled

AMERICAN IRRIGATION FROM ARTESIAN WELLS.
Low Pressure Wells, Southern California.

Irrigation from Artesian Bores.   91

Western and North-western States, it is in the newer and more recently settled sections of the great western country that the most interesting work has been and is being accomplished. To the discovery of artesian water the rapid settlement and reclamation of these great American deserts, comprising the Colorado, Mogave, Death's Valley, Antelope Valley, Armagosa Desert, and the Paramuit Waste, is mainly due.

In the Mogave District, the average rainfall of which is from 2 to 6 inches per annum, the South Pacific Railway Company bored an artesian well, and obtained a good supply. From this start many wells were bored, the land cultivated and improved, and from this section of the Mogave District whose aridity had been so notorious, the wheat harvested in the summer of 1890, took the prize at the county fair over the competitors of all Southern California. It is stated that this district, fifty years ago, was nearly 1,000 miles eastward of its present boundary and many hundred miles to the north.

These are facts worth recording, and I look upon the pessimistic views expressed of the capacity and utility of artesian bore water for the purposes of irrigation by some of the engineers of Australian Government Water Supply and Conservation Departménts, in the same light that such views were held years ago upon the subject of artesian boring itself, as stumbling blocks to healthy progress.

In the last Report of the Conservation of Water Department, New South Wales, there is, however, a " brighter side to the shield."

Mr. J. W. Boultbee, the officer-in-charge, says:—"Artesian water is as a rule suitable for irrigation purposes, and it is only those heavily charged with salt or alkaline matters that are not; and as I can see no reason why this industry, growing daily in importance, should not be an element of immense value, deserving the utmost consideration in developing that northwestern portion of the colony, where the fertility and recuperative powers of the soil are so wonderfully illustrated by the growth of feed after rainfall at the proper season. The average quantity of water required for the irrigation of grain crops, based upon the experience of other countries, may be roughly estimated at 72,600 cubic feet or 543,485 gallons per acre. One inch of rain would equal 3,630 cubic feet or 22,622 gallons per acre. A rainfall of 20 inches would therefore yield 72,600 cubic feet or 543,485 gallons per acre. Six hundred and forty acres would consequently require 46,464,000 cubic feet or 347,830,400 gallons upon them as an equivalent to 20 inches of rain. When it is considered that the flow per diem from the Native Dog Artesian Bore, 45 miles from Bourke, is approximately 2,000,000 gallons per diem or 730,000,000 gallons per year, it will be seen that upon the foregoing basis a supply of water equal to a rainfall of 40 inches per annum per 640 acres is available, or that an area of considerably over 1,280 acres can be supplied with water equalling a rainfall of 20 inches per annum. The cost of the Native Dog bore has been £1,000 4s. 6d. This amount added on to the value of 1,280 acres of land, renders its cost so disproportionate to its value furnished with a water supply—it may be said forever equalling 20 inches of rain per annum—that it seems to me there is a wide opening for the encouragement generally of artesian boring to the great benefit of the colony, and particularly of the North-Western portion of it, where the supposed Cretaceous area embraces a territory of over 28,000,000 acres."

With these calculations of Mr. Boultbee I quite agree and consider he has taken a quite feasible and in no way an optimistic position upon this highly important feature of the utilization of artesian supplies. The plan* that seems the most feasible, and the one that will prove the most satisfactory, is for every farmer to have his own well, and have it so constructed that he will have absolute control of the water. In countries where they irrigate, the rule is, that the water be carried across the country from a reservoir to the land where it is needed; and in this way a great deal of water is lost. Here the reservoir lies under the selection and can be reached without interfering with anyone. By this plan a farmer can select a suitable place for his well which will prove a suitable one for his buildings. If it be found better to expose the water to the action of air and light before applying it to the soil, he can construct his reservoir; he can here grow his own fish; use his water when he likes, without direction or directions from any one; he can use the power to work a feed-mill, thrashing machine, or any other purpose for which he needs power with no extra cost, save the construction of a motor which any carpenter can make at slight expense. These are some of the reasonable possibilities of a proper development of our water resources. In a general way it is believed by those who have had experience in irrigation, that it would be better to collect the water in reservoirs and let it stand for a time exposed to the action of light and heat. In countries where they irrigate, the water they use is mostly

---

*See Report of F. F. B. Coffin, State Engineer, South Dakota, U.S., 1890

Irrigation from Artesian Bores. 93

collected from streams, and has been fertilized by the action of air and heat; it is also charged with silt which adds to its usefulness. The silt we cannot get, but by storing in reservoirs we can get the chemical action of light and air. In a very short time it becomes filled with innumerable animalcula, which is no inconsiderable item. Although invisible they add greatly to the fertilizing effect of the water. The water could also be applied much better when held in large quantities.

Another advantage of this plan would be no more water would be used than would be needed. It is always prudent to be economical in using that which seems to be inexhaustible. Improvidence is not justifiable under any circumstances.

SECTION XIX.

ANALYSIS OF WATER, TEMPERATURE, &C.

IN America analysing water from artesian bores is treated as an important matter, samples being forwarded to, and reported upon by, the Analytical Department of the Government in a most comprehensive and exhaustive manner, the object being not only to determine the suitability of the water for domestic and pastoral purposes, but also for irrigation and the growth of plant life. This practice has been commenced in the Australian colonies, and its importance cannot be overstated. Analyses made of the first water flowing from bores after their completion has proved in many cases to be misleading. This has been owing to the changed and improved quality of the water due to its aeration and settlement after leaving the bore. Water strongly saline on its outflow has been proved in Victoria, and notably in America, to become after a time fit for domestic use and for stock and irrigation.

Under the head of "Permanence of Supplies," and in various sections, I have contended that most artesian water exists in the earth in the form of underground conduits or rivers, the outlet of which is the lowest level of the water-bearing strata, the ocean, and that the fact of artesian water being almost invariably more or less fresh and inappreciably saline on being tapped shows that it has been purified of whatever saline matter it may have derived in its contact with saline strata by the inlet and passage of the large volumes of the fresh rain water, which is the original source of supply. In the same manner I think it is probable that in the course of time, after the continuous outflow of saline or otherwise impure artesian water—which I am of opinion is derived from basin-like water-bearing formations which have no outlet to the sea—the water, by the purifying action of the fresh rain water coming in from the intake to supply the place of that flowing from the bores, may considerably modify, if not entirely correct, the saline or otherwise impure character and render the water in every way serviceable.

Doubts have been expressed respecting the quality of some of the Queensland artesian water, but it has been found in all parts of the colony a good potable water, useful for all economic purposes (including wool-scouring), with the exception of that for producing steam in boilers. The question as to its suitability for vegetation can only be determined by actual experiment, and as irrigation with artesian water is in its initiatory stage in Australia,

NATIVE DOG, BARRINGUN DISTRICT, NEW SOUTH WALES.
DEPTH, 475 FEET; FLOW, 1,700,000 GALLONS DAILY; TEMP., 92 DEG. FAH.

Analysis of Water, Temperature, etc.

(see "Irrigation from Artesian Bores") it cannot yet be predicted with certainty what will be its future. That future will be in all probability a successful one, even if the water be used, as compared to irrigation from rivers, only on a small scale.

A notable feature of some of the artesian supplies is their curative properties. The Agricultural Department of New South Wales reported recently respecting experiments being carried out by the Department in the neighbourhood of artesian bores in the Bourke district that "The water has proved a great success. Vegetables raised by this means are magnificent, and fruits and timber trees are doing well. The bores in the near future will be utilised as curative hot springs. In nearly every case in deep bores the water is warm. The water at the Native Dog bore is delightful to bathe in. For seven or eight months in the year the climate is very pleasant, so that as a sanatorium some of the bores out west are equal to, if not better than, anything of the kind in New Zealand. Already numbers of people suffering from rheumatism and similar complaints have derived much benefit from these waters."

At Dagworth Station, North Gregory, Queensland, the temperature of the artesian water is as high as 197 deg. Fah. Half a mile below the bore an egg may be boiled in the stream, and the water is perfectly fresh and good.

At Woolerina bore, Queensland, the woolwashing is performed ten miles down the creek from the point at which the bore water enters it. The water is hot for nearly a mile and a half from the bore, from which a surface drain has been made. It forms a beautiful clear stream, the water being splendid for drinking purposes.

The following are details of analyses of samples of water from representative bores :—

QUEENSLAND BORES.

	WOOLERINA.	PRAIRIE.	TAMBO.	
Total Solids	38·0	58·0	34·2	Grains per gall.
Silica	2·3	2·4	1·7	,,
Iron and Aluminia	·7	·52	·2	,,
Carbonate of Calcium	trace	3·44	·72	,,
Carbonate of Magnesia	trace...	1·74	·36	,,
Carbonate of Sodium	23·3	not determined		,,
Chloride of Sodium	8·9	34·9	14.01	,,
Chloride of Potassium	1·9	not determined		,,
Sulphates	nil	nil	nil	,,
Poisonous Metals...	nil	nil	nil	,,
Free Ammonia	·70	·04	·30	Parts per gall.
Albuminoid Ammonia	trace	·04	nil	,,

J. BROWNLIE HENDERSON,
Government Analyst.

Artesian Wells:

BOATMAN BORE.*

Total Solid Matter (dried at 220° Fah.),
  34·776 grains per gallon—0·4968 in 1,000 parts.

	GRAINS PER GAL.	IN 1,000 PARTS.
Silica and Silicates	1·232	0·0176
Carbonate of Soda	17·745	·2535
Carbonate of Lime	·889	·0127
Carbonate of Magnesia	trace	trace
Aluminia	trace	trace
Chloride of Sodium	11·123	·1589
Sulphate of Potash	3·165	·0452
	34·154	0·4879
Free Ammonia	·015 per 100,000 parts.	
Organic Ammonia	·002 ,,	,,
Reaction	Slightly Alkaline	
Hardness	4°	

Remarks.—The water is suitable for all domestic use, stock and irrigation purposes. With regard to its use for wool-scouring, it is well suited for that purpose, being a very soft water. No chemicals are required to be added to the water, the soap used being quite sufficient to scour the wool.

(Signed)   H. WOOD,
Under Secretary for Mines and Agriculture,
Sydney.

NEW SOUTH WALES BORES.

YOUNGERRINA BORE.

	GRAINS PER GAL.	PARTS PER 1,000.
Total solid residue	32·984	0·4712
Soluble saline matter	31·892	·4556
Insoluble mineral matter	1·092	·0156
Chlorine	5·100	·0728
Equal to chloride of sodium	8·404	·1200

NATIVE DOG BORE.

	GRAINS PER GAL.	PARTS PER 1,000.
Total solid residue	45·108	0·644
Soluble saline matter	44·044	·6292
Insoluble mineral matter	1·064	·0152
Chlorine	4·500	·0642
Equal to chloride of sodium	7·415	·1059

* This bore is one of the most successful ones in Australia. Depth, 1,500 feet; outflow, 4,200,000 gallons per diem. (See private bores, Queensland) Sample of water sent by the lessee to Sydney for analysis.

Temperature of Artesian Water.

It is generally accepted that the increased temperature of the water is due to the internal heat of the interior of the earth, and increases at the rate of 1 degree Fah. for every 50 to 60 feet in depth. Among the many theories advanced for accounting for the high temperature of artesian water are the following :—

1. Contact with rocks of igneous formation that have retained sufficient heat to regulate that of the water in passing through them.
2. Contact with strata that generate heat by the application of water.
3. The dip of the water-bearing rocks to great depth between the outcrop and the bore.*

The highest temperature in Queensland of which I have information is 197 degrees Fah. (Dagworth Bore). Taking the mean temperature of the air at 60 degrees Fah., and an increase in every 55 feet of 1 degree Fah., we have $195°-60°=135 \times 55 =7,425$ feet as the depth at which the water would derive its temperature of 195.° 'If this theory be a sound one, and it seems to me the most feasible, it affords confidence in making bores to much greater depths than those yet carried out.

---

* The bores greatly varying in depth in the same locality with a comparatively level surface of ground shows that there is great subterranean undulation in the water strata.

Section XX.

MACHINERY.

The first borings for artesian water were made, there is every reason to believe, by the Chinese, and as before observed under the section, "Antiquity of Boring," they were made by the aid of a "trepan," or weight of iron, provided with a set of cutting chisels at the bottom of it, suspended by a rope and worked by a lever by manual power. This method was so slow and laborious that it is recorded the bores took the lifetime of the workers, and in some cases were bequeathed as a legacy to their successors who accomplished them, confirming the well-known plodding industry and persistence of the Mongolian.

The earliest European people who turned their attention to artesian water supply, some four centuries ago, were the French. They used at that time very primitive apparatus, which was very much improved upon in their more recent undertakings, the great wells of Paris some 60 years ago. the great success of which brought artesian water into note, especially in the great farming States of North America.

Amongst the earliest efforts in the States apparatus worked by manual labour was conspicuous, and it is on record that in the further west, up in the territories near the retirement of the aristocratic Indian, where the selectors were few and the land rich, promising, but parched, an old man and his son had taken up some land. They had made a "clearance," built a good log house, with barns, sheds, &c., fenced homestead and yard, and had made, in fact, with one exception, everything snug, safe and complete. The exception was in not having provided a water supply, the periodical want of which, due to little rainfall and dry creeks, was felt to be very great. Borings had been made in the district to hard rock, without finding water, and there stopped. This state of things was not satisfactory to the enterprising spirit of the old man, so after harvest, towards the end of the year, he took a journey to the nearest blacksmith's, invested in a "jumping rig" made after his own heart, including a "jumper chisel" and "sand pump," and went back to his clearing, enlisted his son's help, put up a pine-pole derrick and wooden pulley and spring pole, slung his tool on a rope and went to work. The son's dubiousness when, after they had diligently worked away at the drilling operations month after month through the dreary winter without finding water, would have been fatal to the enterprise of anyone but that of the old man. Some doubt—a modicum—it is related, was confessed to even by him, at a later date, but not divulged at the time, as to his ultimate success. The longed-for possession of good water in the bottom of that well, a supply he would be able to draw upon in winter and summer, led him to pound away at that rock. Once into, he knew, he saw no way out of it but by the lower or other side, and pound it he did, unaided and alone, for more long dreary months, a devotee, almost a martyr, to his self-imposed task, until—*at last*, late one fine evening, a sudden, very sudden roar, an unearthly roar—a rumbling, a rushing of water was heard from below, then up came the chisel-bar, down went the derrick, over went the old man amidst the devastating fury of a solid column of good clear water, rising to the uncontrollable height of 60 feet, flooding everything in the vicinity of its unexpected presence, rushing in its headlong course through homestead, yard and barn, engineering channels for its own particular use, rendering things moist all around, and—it has been running ever since.

Two of the most important wells bored in France are those at Grenelle, completed in 1842, and Passy in 1861. A description of these works showing the time expended in carrying them out, the cost and the kind of machinery used will show the very great advance made to the present perfected machinery, within the last 35 years, in the art of well-boring, and this has been mainly due to its enormous development—primarily in the Petroleum industry—in the United States.

The Grenelle well is remarkable as showing the skill and patience exercised in overcoming the risks and difficulties incidental to well-boring in the earlier modern stages of it. It was commenced in 1832 by an Engineer named Mulot, and took ten years of continuous work before artesian water was struck. This was at a depth of 1,780 feet. At 1,259 feet over 200 feet of the rods accidentally broke loose and fell to the bottom of the bore, and it took fifteen months of experimenting with extracting tools and other devices before they were recovered. This so discouraged the Government that abandonment of the bore was proposed, but in deference to forcibly-expressed scientific advice it was continued until at the depth stated a flow of nearly 900,000 gallons per diem was obtained from a bore of about 8 inches in diameter.

The bored-well at Passy is 1,913 feet in depth, $27\frac{1}{2}$ inches in diameter, and discharges an uninterrupted supply of 5,500,000 gallons per diem. This latter work was entrusted to Kind, a German Engineer, who had sunk many smaller wells in Germany. He commenced operations with a bore of nearly 3 feet 4 inches, and subsequently enlarged it. The tool used was a "trepan" for breaking the rock consisting of two principal pieces—the frame and arms—both of wrought-iron, the framing having at the bottom a series of holes into which cutting chisels were inserted and tightly wedged. These chisels were placed with their cutting edges on the longitudinal axis of the frame, at the extremity of which were formed two heads forged out of the same piece as the body of the tool which also carried two teeth, placed in the same direction, but double their width, in order to render this part of the tool more powerful.

As iron rods from their jarring, and thus fracturing the metal, and on account of their great weight, which would have rendered them unmanageable, and as a great part of the boring was made in water from the upper springs, stout rods of oak 8 inches square were used. The sliding-joint of Eyenhausen—the present "jar" of the English and Americans—was also adopted by means of which the re-acting force of the blows of the trepan were absorbed. After a depth had been reached of over 1,700 feet, the upper part of the boring caved in and filled up the hole. On operations being resumed the bore was contracted to a diameter of 2 feet 4 inches, and at a depth of 1,904 feet a supply of perfectly good water was struck, which quickly increased to 5,500,000 gallons per diem, and it has been running without any diminution whatever ever since. This enormous supply reached a height of nearly 60 feet above the surface. This boring was commenced in 1855, and completed, as stated, in 1861. Its cost was nearly £40,000.

The next method employed in France was the Den sytem, the most important wells bored by it being those at Butte-aux-Cailles to a depth of 2,900 feet, with a diameter of 47 inches; and at the Sugar Refinery at Paris to a depth of 1,570 feet, with a diameter of 19 inches. The apparatus employed consisted of a drilling rod and chisel suspended from the outer end of a working beam or lever made of timber working upon a middle-bearing for a fulcrum, and connected at the other end to a vertical steam cylinder of 10 in. diameter and 39 in. stroke. The steam cylinder was single-acting, being used only to lift the boring-rod with the trepan at each stroke, the rod being lowered again by releasing the steam from the top side of the piston. The construction of the trepan was practically the same as that of the Kind system, the points aimed at being simplicity of construction and repairs, greatest force of blow possible for each unit of striking surface, and freedom from liability to get turned aside and choked. This machine was the forerunner of the English "Mather and Platt" system. The distinctive peculiarities of this latter method consist in the means adopted for giving the percussive and rotary actions to the boring tools, and also in the construction of the tool or boring-head, and of the sand-pump for clearing out the hole after the action of the boring-head. A full description of this machine would show that although it is a powerful apparatus, and entitled to admiration for the ingenuity displayed in the arrangement of its parts, it is not adapted for the peculiar conditions of work in the interior of this country. Not only the complication of its parts, but the great comparative weight of the whole plant and further its excessive cost precludes its use in Australia. Our present simple "American," an equally powerful and effective apparatus for deep drilling, with its record already achieved in Australia for some of the deepest drilling in the world, may be taken as symbolical of the requirements of the physical conditions of this country compared to those of Great Britain. Lightness, simplicity and power combined are the ruling desiderata for this country.

In the early modern stages of artesian water development the diamond drill was brought into requisition, and is being now employed in the exceptionally hard strata common to auriferous gold-mining country—notably in Western Australia, in the search for artesian water.

The difference between the hardness of some of the rocks and the best of steel cutting tools, is in favour of the former. This gave rise to the adoption of the hardest known substance—the diamond—to perform the work of boring through them. Fortunately this substance existed in another form besides that of the gem, the costliness of which would have precluded its use in boring operations. The imperfectly crystallised form known as "Carbonate" which, until the advent of this drill, had little commercial value, has been admirably suited for the purpose. It is of a dull black colour, unlike the gem variety, has little cleavage—a singularly valuable quality for boring purposes—as the liability to split is thus removed. It is, however, by abrasion alone that the diamond can be worked, and not by percussion, as in steel-cutting tools. It is a well-known fact that if two substances, one harder than the other, are rubbed together, the surface of the softer one is worn down by friction. It was therefore decided that the action of abrasion should be employed by means of the diamond for the purpose of piercing the harder rocks. It would appear at first sight that this process is a slow one, but with a due degree of velocity given to the turning of the tool it is, in fact, much more rapid than the ordinary method of breaking the rock by percussive or falling tools. The diamond drill is capable of piercing the hardest rocks; even emery has been bored at the rate of two inches per minute. The rate of progress during actual boring is from two to three inches per minute. In soft strata such as clay, sand and other alluvial deposits this drill is worse than useless. When these strata are met with other tools must be employed until the hardest rocks are again met with. It must also be observed that the borings made by this machine are of a comparatively small diameter.

The next consideration in connection with this subject is that of the best mechanical means at our disposal for accomplishing the purpose in hand—obtaining artesian water.

The geological features of the subject, however scientifically and fully they may be treated, however able an account may be given of the physical characteristics of the country, can be at this early stage of the development of artesian water supplies in Australia at the best no more nor less than speculative generalisations. Geologists will, I know, freely admit that they are dependent upon the exposition of the strata given by actual borings for the most reliable data upon which to frame their reports and finally perfect their maps. If the water which undoubtedly lies in the ground—already proved by the drill—be worth the getting, the very best means must be adopted in order to give it vent to the surface; and this consideration of the subject, although it may appear to some people to be a trivial one, lies within the special province of a class of men—the engineers—who, I submit, have filled, and are filling, probably the foremost position in the material development of these colonies, and who, having had to directly interest themselves in this matter, know it to be a very important one.

The following description of this machinery as used in the great petroleum fields of America was published by me in the *Leader* newspaper, Melbourne, 1879, and also in the first edition of this book, Brisbane, 1883:—

"Having made all the necessary arrangements, we will now commence operations and accompany the driller in his downward journey through earth, shale and rock. The engine is fired up and the attachments made giving the crank motion to the working beam, which in turn moves the cable and the drilling apparatus. The driller takes his seat on a high stool above

Machinery.

the chosen spot, adjusts the centre-bit or drill with great care, and down it goes through the conductor pipe, striking from thirty to forty blows per minute. Between the strokes the tools require to be moved around to make the aperture uniform and to prevent them from wedging fast. With this is also connected a slight downward motion every few strokes by a turn of the temper-screw. The first operation tells the entire story. Day after day, night after night, the drill is kept moving up and down, cutting from one to six inches or even twelve inches of hard rock and shale per hour, according to hardness. At intervals the centre-bit is drawn up badly worn and battered, and a rimer let down to enlarge the hole and put in the finishing touches, by making it smooth and round, and these are followed by the sand-pump to raise the debris or excavated material."

The above is a popular description of machinery used by the drillers of Pennsylvania, and the lessons learnt by these early American miners were by the force of hard practical experience the safest, surest and only true guide to a perfect knowledge of the business. Mining Engineers who had graduated in the best schools of the old world and the new came in the subsequent years of the development, but their plans were of too elaborate and complicated a nature to admit of permanent adoption. None of the companies, much less individual miners, cared to indulge in such costly and complicated machinery as that in use in other branches of mining. They have, after trial of other machines, adopted the rope or pole and the "jumper" chisel, with such other appliances as combined adequate strength with simplicity.

An immense number of devices and inventions have been fairly tried in this highly practical and experienced community of Pennsylvania, intended for drilling oil wells without sufficient success to ensure even their partial adoption.

In America the drilling operations for petroleum set the earliest lessons in sinking for artesian water, followed at a later date by Canada and other parts of the United States, for coal oil and also water, and was the immediate forerunner of the drilling plants now adopted in Australia. This mode is that by which the desiderata of power, speed and simplicity of construction and working are fully combined. The Continental European modes, from their extreme elaborateness, complication of parts and comparative slowness in working, are not likely to find favour here. The question then becomes narrowed down to the acceptance of what is known as the American, or Canadian, system; but as the partisans in America and elsewhere of that

system are upon an important matter of detail, split into two camps, all that remains to be considered is which form of the American system is the best adapted for Australian use. The difference of detail consists in the use of a cable for drilling in one mode, and the use of poles for the same purpose in the other.

If a cable be used it is necessary to have a large winding drum with much heavier, costlier, and more complicated machinery than when poles are used. Under the pole method stout poles of wood in long lengths of from 16 to 20 feet with screwed wrought-iron ends are used in lieu of a cable. The cable has been largely used, as it is—so far as the lowering and winding-up of the drills are concerned—extremely simple in working. Where simplicity can be gained without corresponding disadvantages it is well to favour it, but where a manifest inferiority exists, to choose simplicity in opposition to complexity, for its own sake alone, is absurd. To the cable plan serious drawbacks occur, one of which is that the bore-hole is apt to become crooked, necessitating expensive riming or straightening before the necessary tubing for protecting the sides of the bore can be inserted. Another is that the weight of the drilling tools, when at work, stretching the cable until they touch the bottom and bounding from it give a quick and rebounding blow. This is called, in America, "bouncing" the drill, and is very destructive to cables, jars and band wheel shafts. This mode of drilling by impact alone, although assisted by a twisting action of the cable, will never, I am of opinion, become general for artesian or deep boring. In rocky strata for limited depths, or in places where the straightness of the bore is of little moment, the latter may, however, be applied. In bores of great depth the pole system requires—as will be readily understood—exceptionally powerful lifting apparatus, engine, &c.*

The plant required for making an artesian bore by the pole system is as follows :—

The Derrick.—This may be constructed of timber or of angle and flat iron. It stands directly over the bore, and for a *full-sized rig* of timber is 20ft. square at the base and 72ft. high, the four corners converging so as to form a square at the top 3ft. in diameter, upon which rests a heavy framework for the reception of the crown pulley, over which the rope for lifting the drill and

---

* In the Malvern Hills bore, Queensland, the second deepest in Australia, and one of the deepest in the world, the contractor found the pole system impracticable, with his plant, at a greater depth than 3650 feet, and a 2in. Manilla rope cable was then employed to a depth of 3948 feet.

poles plays. There are six foundation posts, each 18in. square, two derrick sills 21ft. long and 10in. square. On the top of these are laid floor sills 8in. by 10in. and 20ft. long. Then follow floor sills and floor planks on top with a hole in centre for the drill to pass through. The derrick is then built up to the required height with planking, if of timber to the required height with planking returned for the four corners, and diagonally and horizontally braced.

Samson Post.—This is made of timber 18in. by 20in. at the bottom, and 18in. square at the top, and 13ft. high, dovetailed into the main sill, held by fitted keys and braced by braces.

Walking Beam.—This is of hard wood, 26ft. long, 12in. by 26in. at its middle, where it rests on the saddle or samson post, and is bevelled on its lower side to 12in. square at its ends. At its front end a slot is cut, in which the drilling hook is hung.

Bandwheel.—This is 9ft. in diameter. It is connected with the pulley of the engine by a belt. It communicates power to all the movable parts of the apparatus. It is fastened upon a shaft, which has at one end a crank or arm with several holes, in either of which—according to the length of stroke of the walking beam desired — is inserted a wrist pin, to which is attached when necessary the "pitman," which is connected by a stirrup with the walking beam. The rotation of the band wheel therefore causes a rocking motion of the walking beam. There is a knuckle joint, in which a sand-pump reel works.

Engine.—This is generally a portable one of not less than 15 horse-power, having a cylinder at least 9in. by 12in.; although smaller engines, in good hands and under favourable circumstances, can be used. The engine is securely anchored on the ground and by means of its driving pulley and belt motion is communicated to the band wheel, and through it to all other parts of the machinery. The motion of the engine is under the control of the driller in the derrick, for the throttle valve of the engine has a large grooved wheel attached to it, and from this grooved wheel the endless cord—called the "telegraph"—extends to the derrick and passes round another grooved wheel, which is fastened to a post convenient to the driller.

Drilling Tools.—These consist of a sinker bar 12ft. long, weighing 400lb.; jars 6ft. long, weighing 300lb.; auger stem 32ft. long, weighing 1050lb.; bit, 3ft. 4in. long, weighing 140lb. These are screwed together, forming the "string" of drilling tools. There is also a rimer 3ft. 4in. long, weighing 160lbs., for making the hole round. The sinker bar and auger stem are made of 3½in. iron ; the jars have a play of 9in., and are 5½in. in diameter. A pair of wrenches 3½ft. long, and weighing 75lb. each. A temper screw for paying out the tools and rods as the drilling gets deeper. The joints of the rods and tools are all made with a taper joint, which is very superior to the ordinary joint. Its pin and box being tapering, more than double the strength is added at the base of the pin, which is the place where strength is required. As the pin enters half-way into the box before the threads engage, it takes but half the number of turns to screw a joint together that is required in the ordinary joint. As the bit must be unscrewed and replaced several hundred times in drilling each well, the saving of labour and time is very considerable.

The Sinker Bar.—This gives force to the blow upward when necessary to "jar," or loosen the bit when fast.

The Jars.—These are made in two parts, and are like two long links of a chain. Both parts are slotted, and the cross-head of one passes through the slot of the other. When extended the jars are 6ft. long, and when closed 5ft. 3in. The difference, 9in., is the play of the jars. The function of the jars is to give a blow upward. If the auger stem were attached directly to the sinker bar it might stick in the rock so that it could not be drawn out, but an upward blow will jar it loose, and this blow the jars give.

The Temper Screw.—This screw connects the walking beam and the cable, and it is gradually let out as the bit cuts into the rock. It is hung on the drilling hook attached to the well end of the walking beam. The screw is 5½ft. long and 1½in. in diameter, with a square thread of two threads to the inch.

Bailers, or Sand Pumps.—Of these there are several kinds. They are made of wrought iron 20ft. long. One has a bail at the top and a foot valve at the bottom. The valve stem projects downwards below the pump, so that the valve may be opened and the contents of the pump discharged by resting the stem on the ground. Another is 5ft. long; it has a suction valve on the end of a plunger. The sand pump line is fastened in an eye at the top; when the pump stops at the bottom of the well the plunger descends to the bottom of the pump. The leather bucket is so constructed as to go down in the pump readily, but on pulling back it flattens, and becomes a tight piston. When the line is pulled up, the valve and plunger are pulled through the pump, sucking in the debris from the bottom; the lower valve closes and retains the contents. If it should get fast in the well, it can be "jarred" out.

The Poles.—These are made of tough light timber (preferably of Queensland yellow-wood or Crow's ash) of a circular form,

3in. or 4in. in diameter, and in 16ft. to 20ft. lengths. They have wrought-iron taper-screwed ends for coupling and uncoupling. By using the "jar" below these poles the chisel, or bit, alone strikes the rock, the poles above it falling a less distance than the play or space between the upper and lower links of the jar. Previous to the use of this arrangement the weight of the poles at great depths was an evil of serious importance, for when percussive motion was given to them they vibrated with great force, and, striking against the sides of the bore, detached portions of loose rock which fell upon the top of the working chisel. As the bore is generally full of water from the upper springs, large light, but strong, wooden poles lose a great deal of their weight in the water, which is a desirable point, and the vibration, owing to their volume, is reduced to a minimum. Another important advantage in the use of the poles is that, from their rigidity, they admit of certain and trustworthy turning of the chisel at its work in the bottom of the bore-hole, a most important matter in the quick and safe execution of the work. The height of the derrick admits of three lengths of rods being lifted at a time without unscrewing them. A large size drive-pipe is first driven down to the rock; the well is then drilled down below the lowest top water seams. Tubing of a size according to the probable depth of the bore is then inserted, and adjusted water-tight upon the rock. A smaller bit is then used, and the well drilled that size as far as desired.

Tubing.—See Section XXII., " Tubing, and Durability of."

Torpedoes.—After a well is sunk to its proper depth it is sometimes torpedoed.* A torpedo is a charge of nitro-glycerine or other explosive in a suitable shell, which is lowered to the water-bearing rock and exploded with the effect of enlarging the bore at that point and opening fissures into the surrounding rock. A shell is suspended by a cord on which runs a hollow weight, and the cord guides this to the firing head, which is constructed so as to communicate a blow to a percussion cap and explode it. The detonation will explode the nitroglycerine.

Commencing the Well.—When the walking beam is mounted the drilling hook is hung. If the earth is firm, a hole 12in. in diameter is "spudded"; a wooden conductor made in octagon form is then inserted in the rock. This prevents the surface earth from caving into the well. If, however, the rock is at such a distance below the surface, or if the surface soil is of

---
* This is only resorted to in low-pressure wells in which the underflow of water is sluggish.

such a nature that it is not practicable to "spud" a hole for the conductor, the rock is reached by a drive-pipe of the diameter determined upon armed at its lower end with a steel shoe. This is driven through the surface soil by means of a heavy maul made of a sound log of timber 15ft. long, similar to the manner in which piles are driven by a pile driver. The pipe while being driven has a heavy iron cap to prevent injury from blows of the maul. The maul works by means of guides between vertical planks set up temporarily, and is lifted through the crank movement by the engine. In case a boulder or other large rock is encountered before the rock is reached, it is drilled through by the drilling tools, and the drive pipe can be afterwards driven through the hole thus formed. By means of a special bit or "undercutting tool" the hole may be enlarged below the drive-pipe. When the rock is reached the hole is drilled into it a few inches, and the drive pipe firmly driven in so as to form a tight union, in order that no surface water can leak or dirt get into the well. As soon as the hole is deep enough to "bury" the tools, the string of tools is let down, connection is made with the walking beam, the engine started and drilling commenced, the tools rising and falling with each rotation of the crank. When some 12in. of rock have been cut the drill and rods are raised, swung out of the way of the sand pump, which is then run down. The bit is unscrewed and is dressed or sharpened, carefully cut to the size of the gauge, and one is screwed on ready to drill again.

A process of sinking, to overcome the difficulties experienced in getting down the tubing in quicksand (the drillers' bugbear) is now before the notice of pastoralists in Australia. It is an extension of the hydraulic system of sinking which has been used for many years,* and which consists of working tubing which is revolved, and which has a cutter head screwed on to the bottom of the tubing for cutting its way. Inside this tubing a pipe of small diameter is lowered to the level of the cutter head, and through it water is forced by a steam pump on the surface, the pressure of the jet being sufficient to force the loose sand up the tubing and discharge it at the surface, the tubing going down on the removal of the sand by its own weight, or with additional pressure. The process consists also in the use of an expansion drill for cutting through boulders, gravel or rock, working below the bottom of the tubing by a lift and drop movement, the cut material being forced, as was the sand,

---

* The author reported in 1886 for the Railway Department of Queensland, on an hydraulic boring made at Muckadilla.

up inside the tubing to the surface. The very obvious difficulty in the use of this mode—supposing its efficiency to be as great as is claimed for it—is in the large quantity of water required for the force pump. Water is valuable in Australia especially in proportion to its scarcity. The item of cost of haulage of water long distances, for the engine alone for boring plants and for camp use, is a serious matter, which is only obviated on water being struck in the bore.

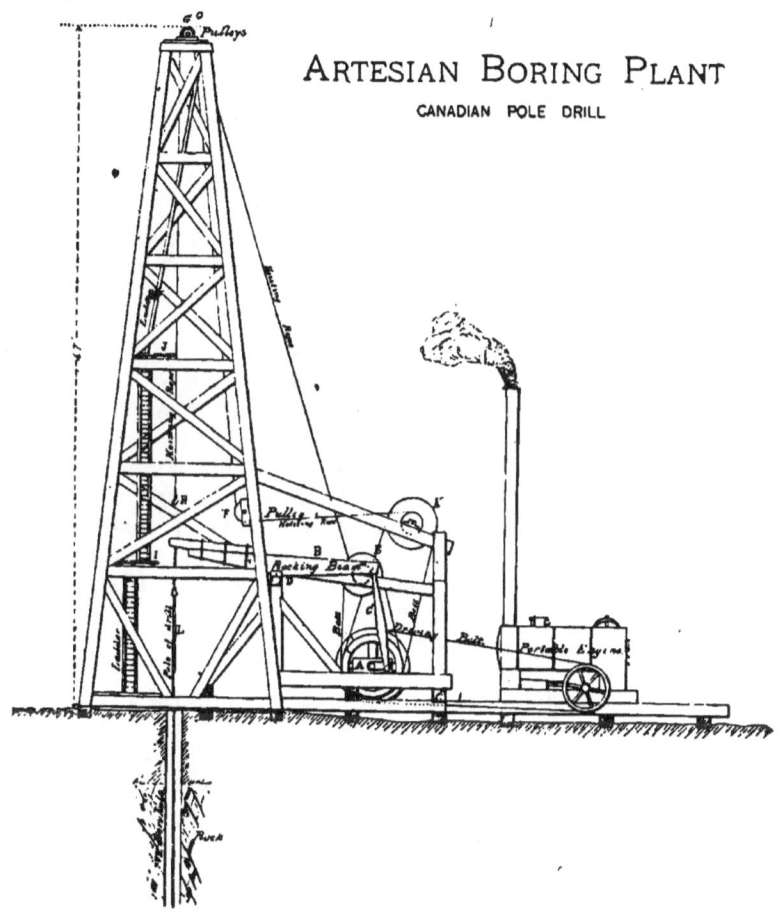

ARTESIAN BORING PLANT
CANADIAN POLE DRILL

The machinery now almost exclusively used for deep drilling in America and Australia will be best understood by reference to diagram. A belt from the engine drives the pulley and shaft at A, with a crank for working the end of the rocking beam B to which it is connected by the connecting rod C. The rocking beam B is pivoted and supported at the centre D, the end over the bore hole having the drilling poles attached to it. The winding drum is at E beside the rocking beam, and is driven by a belt from the first shaft at A. A wire rope from the winding drum passes over the pulley G at the top of the derrick ending in the hook H immediately over the bore-hole. The object of this arrangement is to raise and lower the different lengths of boring poles also the tubing for lining the bore. The sand pump for removing the excavated material is worked by means of another wire rope long enough to reach the bottom of the bore, however deep it may be. This rope is coiled upon drum K and passes under pulley F and over pulley O, which is alongside G on top of the derrick.

Ladders and two stagings, I and J, are provided for the workmen to attach and detach the lengths of drill poles or tubing as they are hoisted up. The cost of a complete rig with set of blacksmith's tools and poles (see Appendix C) to drill 2000 ft. is from £1250 to £1600 according to requirements. The whole rig takes to pieces for facility of transport. The men required for working it are one engine driver, one driller and one derrick man, if near a township; but if in the far away bush a cook and blacksmith are also required, making in all five men. The Canadian drillers work twelve hours a day, and when a double set of men is employed the work goes on day and night. Working twenty-four hours a day the rate of progress is from 5 ft. to 100 ft. per day, according to the class of rock or clay passed through. The cost per foot increases for great depths.

The bore is lined with tubes to prevent the sides from falling in where the strata passed through consists of soft or loose material, and also to shut off brackish water sometimes found near the surface. The tubes are also necessary to carry the water to the surface and prevent it escaping into the dry ground between. The tubes used (see Section, "Tubing") are of wrought iron screwed together, with "swelled joints" to admit of deeper screwing, and are "flush" inside—that is, free from any projections which would interfere with the free passage down and up of the working tools.

In the earliest American bores of moderate depth—up to 600 or 800 feet—a cable was used in lieu of the Canadian "pole system," poles being used for drilling to a greater depth, and at extraordinary depths the cable was again reverted to. The practice now in America is to use both a cable and poles alternately in the same bore, the length and position of each being decided upon by the nature of the strata and the depth of the boring. This is also the practice with some Australian drilling contractors, each one it need scarcely be remarked, exercising his own particular views upon this detail in the working of the main principle of the system, which is essentially the substitution of a balanced beam (as shown in the diagram) resting upon a middle bearing to one end of which the cable or poles with the chisel attached are suspended, the other end being given an upward and downward movement by means of a crank movement worked by a steam engine in place of a rotary movement as in the diamond drill and other rotary (boring) machines.

Among the many inventions of earth-boring machines and tools (the records of the Patent Offices, particularly of America, show that they form a legion, most of them being highly ingenious but impracticable) there is a combination of the pole and diamond drill systems, which has been recently brought forward in Australia from Germany as a new idea, although it was reported upon by me some eighteen years ago. In one of the late Government Reports* is the following :—" The boring plant in general use for sinking artesian wells does not fulfil the conditions of a perfectly scientific apparatus, because it does not produce a core by means of which the nature of the strata and the angle and direction of the dip can be fully ascertained." This combination of the diamond drill and cable, or pole system, for the purpose of general work in the interior of this country is, I think, out of the question. The strata of very hard ground never exceeds a slight thickness, and although it is tedious work, where met with, to drill through it, even with the best of steel tools (sometimes requiring a week or more to penetrate even eight inches) it is preferable to the undertaking being burthened with the extra cost of additional plant, extra haulage, and the setting up and taking down the auxiliary "combination" apparatus of a diamond drill, with its extra line of special diamond drill tubing. It is not so much a question of feasibility in a mechanical sense as in an economical one. I do not think, therefore, that this "combination" will be generally adopted. There is the diamond drill or the "Calyx" for obtaining cores in boring for minerals, and the former is being extensively used for that purpose, and although no actual compact cores are produced by the artesian machinery now in use samples of the strata drilled through are, when forwarded to the Geological Departments, readily assigned to their particular class of rocks.

---

* Queensland Water Supply Department, 1893.

The present machinery for deep artesian drilling is, I am of opinion, all but perfect. Like the locomotive it may be open to improvements in minor details which extended practice may call for, but I cannot conceive that for the general purposes of its special work in the interior of this country it is likely to be superseded by any other mode. It has a unique history, and in its embodiment of the best thought, exceptionally thorough and extended practical experience, and persistent energy of a long succession of practitioners, it stands, I believe, as near perfection for its purpose as human efforts can make a machine.

In the management of these machines, from the commencement of an artesian water supply in Australia to the present time, the services have been enlisted of Americans, mostly Canadians from British North America. That they have since the time (but a few years ago) when the colonies decided upon obtaining artesian water, effectually carried out the business to a very successful issue goes without saying, and the thorough and workmanlike manner in which the art of deep well boring has been in a short time placed permanently before Australians, to their enhancement, by these exponents of it, and the great results achieved by them call, I submit, for unqualified praise and consideration.

DRILLING AT MARATHON STATION, QUEENSLAND

MANWELTON No. 2, QUEENSLAND.
1,474 FEET DEEP; 1,200,000 GALLONS PER DIEM.

SAXBY DOWNS No. 2, QUEENSLAND.

SECTION XXI.

MECHANICAL POWER DERIVABLE FROM ARTESIAN BORES, AND TOWN SUPPLIES.

THE mechanical power derivable from the pressure given in the outflow from artesian wells, although it varies in different bores, is a constant economical mechanical asset, the value of which, even now, is little understood in Australia, and the prospective value of which, when the bores become multiplied over the face of the country, can scarcely be estimated or realized. The power is both ubiquitous and unique. It is cropping up in out of the way inland places where steam power is not feasible and ordinary water-power out of the question. The power is direct and probably one of the most economical conceivable. To meet the requirements of the numberless operations from running a sewing machine or a cream separator to a saw or flour mill and for extinguishing fires it is most desirable. By the application of the power to motors, such as the Turbine and Pelton wheel, the possessor of an artesian outflow may be held to be fortunate in the extreme.

The power being free from working expenses in its production must ensure a large use of it in the near future. It need hardly be said that it is largely availed of in America for the purposes mentioned, and also for electric lighting in towns it is from its steady and reliable nature in great demand. It is also largely used in that country for sheep-shearing purposes.

At the city of Aberdeen (U.S.A.) an artesian bore was sunk in 1882, and the flow from it demonstrated very successfully its efficiency in forcing the water through the street mains for the supply of the town and for fire protection. The first application of this power was also made in driving machinery by the Electric Light Co. of that city, and at the present time scores of motors are in use for working printing presses, elevators, small mills, &c. At this same city of Aberdeen, owing to the level character of the surface, it was difficult to get drainage for a sewerage system. By the advice of B. Williams, Civil Engineer of Chicago, the sewerage was conducted to the outskirts of the city into a deep deposit chamber. An artesian well was then sunk in the pump house. From the top of the well the water is conducted through four-inch pipes to the pumps, which raise the sewerage of the city from the deposit chamber, through a height of 23 ft., at a rate of over 2,500,000 gallons per day. All this by the direct pressure of an artesian well that cost £680; was sunk in less than 90 days; works automatically; requires no other attention than that of one man to oil the pumps; requires no special building; needs no repairs, and costs nothing to maintain.

The power is also being utilized in South Dakota for working a flour mill in which, with a small bore of 4 in., from fifty to sixty casks of flour are ground per diem.

It is gratifying to know that this power is already being used in the water supply by reticulation of an inland Australian town—Thargomindah, Queensland, the example of which is being followed by other inland towns. To Thargomindah belongs the honour of having first applied a scheme for the general utilisation of the bore water. For some time there has been before the people there a proposal to lay the water on to the dwellings. This idea has now partially assumed a definite form, it having been decided to reticulate the town with water mains. The application of water-power to the existing electric lighting machinery may come later on.

The bore is located about three-quarters of a mile from the town, and such water as has hitherto been used by the householders has found its way down the water-channels, and has thus been fit only for washing-up purposes.

The main will run from the bore to Dowling Street, intersecting Sam Street (the other principal thoroughfare). From the points of intersection of both streets, pipes will branch east and west to a sufficient distance to take in all the dwellings. The distance the water will have to travel will be rather an advantage in the case of Thargomindah, since the water when it issues from the bore is at a temperature of 166° Fah., and, consequently, requires considerable cooling before it is usable. It is, nevertheless, of excellent quality, and the laying of it on to the houses will be an inestimable boon. At the present time the fire insurance rate averages something like 45s. per cent., and a reduction of this heavy tax is not unnaturally looked for.

The example thus set is one which will doubtless be followed in both Charleville and Cunnamulla—other Maranoa district towns; in the latter place particularly, for there the artesian water is literally wasted. The estimated cost of the work at Thargomindah is £2500, of which the Bulloo Board finds immediately £1100. The remainder (£1400) has been obtained as a loan from the Government on terms extending over forty years. The bore has a depth of 2650 ft., and yields a flow of

670,000 gallons per diem, the water-pressure being as high as 230 lbs. to the square inch.

In the report of the Queensland Water Supply Department (1894) the following is given as the power of the Government bores—information from the private bores not being available. The power given is based on the static pressure and a mechanical efficiency of 8·5 per cent. of a Pelton wheel:—Barcaldine bore, 1·25 horse-power; Saltern, 0·27; Blackall, 8·04; Tambo, 1·07; Cunnamulla, 41·59; Charleville, 123·41; Thargomindah, 63·51; Winton, 3' 70".

Section XXII.

TUBING, AND DURABILITY OF.

THE tubing, or "casing" of the Americans,* for wells has, like the machinery, been brought to its present state of perfection by a slow process of experience and the advance in manufacture.

The bored palm logs of the Arabs (alluded to in the Section, "Algeria") and similar contrivances used by the Chinese were probably the earliest efforts made to sustain the sides of the bores in loose ground.

In England and other parts of Great Britain—before the advent of the Waterworks Companies—when the water was obtained by well-making and "pumps" were almost universally used, bored oak logs were used. Cast iron and even leaden tubes were afterwards adopted until the introduction of wrought iron welded tubes which took their place.

In the early sub-artesian boring operations in Australia rivetted sheet iron casing was used sometimes of a double thickness. This tubing gave great dissatisfaction, and is accountable for a great deal of extra expense in putting down the earlier bores. There was, however, at that time, on account of the enormously high price of wrought iron tubing,† scarcely any alternative than the use of it.

The kind of artesian tubing now used in Australia, and which as regards make and, in the best brands, quality of iron, is a really valuable adjunct to the operation of obtaining artesian water, has been largely used in Russia, Roumania, Italy, Austria and on the Caspian Sea for oil wells for many years, the demand for which has been mainly supplied by various British makers. Well borers and contractors evince their preference for various brands of tubes, and there is, in point of fact, no brand on the market at the present time that can claim any special superior make or quality of iron, nor is there any that I am aware of that holds a prohibitive patent right to any particular make.

The best tubes are those of so superior a quality of iron and make that they will stand all the tests specified by the most

---
* I prefer the term tubing because this protection to the bore is tubular. A casing may be square or any other shape.

† I paid at this time eight shillings per foot in Brisbane for six-inch wrought iron tubing, at the same time urging importers to lay down a large stock, with no result.

exacting engineer, and at the same time be turned out of a minimum thickness, but which admit of efficient screwing, thus reducing the weight for freight and in handling and using at the work. The tests imposed by some of the Australian Governments are compression, tensile or pulling strain, lateral or side strain, and twisting strain. If the iron be of inferior quality, however perfect in shape the tubes may be, the screwed part will give way—the thread "drawing"—and may be thus ruin the bore. The whole success of a bore depends upon its tubing; costly accidents, sometimes irreparable, have occurred, causing the total loss of a bore from defective tubing.

DURABILITY OF TUBING.

The question as to the life of artesian wells, *i.e.*, the durability of the tubing, is one that cannot be disregarded in a work of this kind, and although it may be thought by some to raise an unnecessary point in the subject of artesian supplies and to give cause for anxiety to others in undertaking boring operations, the value of the water is so great that even if, in extreme cases, the tubing did not last many years, it would pay to put down another bore on the site, always supposing nothing could be done to repair the existing bore.

There is a great deal of interesting controversy going on on this subject. It is manifestly only a question of time when the iron tubing will be destroyed by corrosion. The time will vary in different localities; that is corrosion will go on faster in a district where the water is strongly impregnated with corrosive agents. This action is dependent not only on those agents, but on the nature of the iron itself. If this be of an inferior quality, although of great thickness, it will corrode faster than a tougher superior iron, although the latter be less in thickness.

The result of actual experience in England of tubing used in artesian wells which have been made over sixty years is that no appreciable corrosion has taken place. In some wells, however, corrosion has had its way, and the remedy has been to line the bores with copper or brass tubes. Seeing that nearly all the Australian bores strike fresh first waters and that the deeper artesian water is also of good quality and free from saline or other injurious elements, it may, I think, be safely concluded that corrosion will not take place to any appreciably detrimental extent for a long time. The effect of corrosion, or rusting, is that the sand or soft rock or other debris which has settled outside the tubes becomes oxidised and *hardened*, thus tending to support the corroding skin of the tubes—a matter of some moment.

As to re-tubing the bores ; if the bore has been properly made and tubed, it would, I think, be practicable to re-tube it when necessary. I think it is probable that, as a result of experiments now being made by metallurgists, a mode of rendering the tubing rust-proof may be established.

Section XXIII.

SUB-ARTESIAN OR SHALLOW-BORING.

THE late and progressive development of the enormous underground supplies of artesian water which extend over large areas of country in these colonies, particularly in New South Wales, South Australia and Queensland, has given an impetus to well-boring of all kinds.

There can be no doubt now in the minds of the most dubious that there is an immense accumulation of water in the rocks and sands of the upper crust of the earth, and although this is now being obtained, with very pronounced success, in large self-discharging quantities from the deep, or artesian, sources of supply—ranging from 500 ft. to thousands of feet in depth—very considerable quantities of valuable water may also be obtained in many localities at shallower depths, which if it does not reach the surface may be brought there at a small cost.*

The following is a description of actual work performed under my own superintendence and its results in obtaining a "sub-artesian" supply, this supply being fully defined under the head of " Definition of Artesian Water."

The country operated upon—and it may fairly be taken as a representative Australian one—was that of the head of the Barcoo,† Birkhead Creek, 25 miles above Tambo, Central Queensland, on the southern side of the ranges running west, and consists of Rolling Downs, through which the usual creeks—which are dry the greater portion of the year—have made their way. The geological formation (which may be taken as entirely reliable in regard to water-bearing) consists of sandstone dipping from the ranges to south and west. The "weathering" or disintegration of the ranges during a long period of time has formed, by the action of running water during the same period, large deposits of sand in great wide flat stretches of country which were formerly valleys between successive ridges, now showing above the sandy flats, the lower part of which has been turned into porous sandstone, overlaid by a loose sand deposit with seams of clay varying from 15 ft. to 20 ft. in thickness. It was in the porous

---

* See Section, "American bores," page 16.

† The Birkhead Creek, Central Queensland, is that mentioned in Mr. Jack's Paper (see section " Queensland ") as one of the Blythesdale braystone areas. A number of successful bores were made by me in this " Rolling Downs" formation many years ago in various parts of Queensland.

water-bearing sandstone that the large supplies of this "shallow" or sub-artesian water have been obtained. "Soakage" or first water was met with in the loose sand deposits at depths of from 6 ft. to 33 ft. This was entirely shut off by water-tight artesian tubing, and the second or permanent water was struck in the porous sandstone at depths of from 17 ft. to 90 ft. Drilling was continued up to 120 ft., thus ensuring a full supply. A four-inch cylinder pump (the casing used being six inches inside diameter) was used at each bore for testing the supply, the tests being made after a long period of dry weather. After continuous pumping for hours the level was not lowered, but on the contrary, the supply was increased as the pumping opened out the water seams in the porous rock, and thus increased the flow. The capacity of the pumps when worked as they are at a great number of similar bores on the Darling Downs—by windmill power—is at the lowest discharge 15,000 gallons per diem. The pump, worked by two men, threw 800 gallons per hour at these Barcoo bores.

A belief prevailed in the district that no "shallow" water—*i.e.*, within 100 ft. in depth—could be aught else than "soakage" from the immediate surface, and that during droughts it would yield no reliable supply. This belief was, however, very successfully combated, the facts of the case being that the first or soakage water stood between 6 ft. to 30 ft. below the surface. This water was entirely shut off by water-tight artesian tubing, the bores being perfectly dry until the second or permanent water lying in the porous sandstone was struck at from 17 ft. to 100 ft. The level of these waters has not altered during many months of dry weather, nor during or since heavy floods. If there were any connection whatever between the soakage water and the second water, the latter would rise to the former. The fact that it has not so risen is conclusive proof that the two waters are distinct. Moreover, the top or soakage water lies in loose fine silt and drift. Wherever this water could run it would carry this silt and fine sand with it, and if it had found its way into a sump or bore hole below its level that hole would soon—in a few hours at the outside—be silted up. That these holes did not silt up, but that each bore plumbed the same, to an inch, some four months after, as it did at its completion, is a further proof that the lower water comes out of the porous sandstone lying below the level of the bottom of the tubing, which tubing was inserted to shut out the loose sand and drift. The tubing was forced down through the stiff clay seams lying below the sand deposit and into the top of the sandstone.

Sub-Artesian or Shallow-Boring. 121

In another boring at Cooper's Plains, South Coast Railway, Queensland, perfectly fresh water was struck at a depth of 82ft. only, the water rising to within eight feet of the surface. The strata passed through were sandstone rock, carbonaceous shale, and some 17ft. of compact, tenacious brown clay, at the bottom of which in a sand drift the water was struck. The water-tight clay prevented all possibility of the water being simply soakage. The water had come from a long distance to the site of the boring, and the height to which it suddenly rose indicated very considerable pressure, which was undoubtedly due to a large body or run of underground water existing continuously far away to the Ranges or higher lands of the Downs.

Previous to making these bores shafts had been sunk in the beds of and near creeks, and were failures, as on getting to the first or "soakage" water the very strong loose sand drifts—which artesian tubing effectually shuts off—could not be got through by a slabbed shaft.

WORKING APPARATUS.

The question of the best mechanical means of getting at this sub-artesian water supply—this water at moderate depths—(it is to be understood deep artesian machinery is not now under consideration) is one which practical men—those who have had an opportunity of thoroughly studying the subject—have for some time set at rest. In the year 1878 the Victorian Government started one of the first boring machines used in Australia. This was placed under my superintendence. Since that time various types of machines of English, Australian, American and German make have been used by the Governments and landholders of the various colonies, the Queensland Government having first employed a water-boring machine—also under my superintendence—in 1882. Each of these machines has given more or less satisfaction, but there can be little doubt that neither of them has proved an unqualified success. Complication of, and too many, parts; a lack of simplicity in the movements, both for boring in alluvial ground and drilling in rock; liability in transit and working to breakages, or to some hitch or other in the complicated wrought and cast iron mechanism, with its shafts, wheels, pulleys, chains, spiral and bevelled screws, levers, and all the parts necessary to produce that complex movement which has its origin in a very simple application of the original power—which must in any case be applied —have produced this result. In the consideration of this phase of the subject it will, of course, be understood that no engineer, or any other reasonable person, would desire to depreciate the inventive skill shown in these various machines. They are undoubtedly a means to an end, but they have failed in adaptability to the conditions of the bush—the interior of this country.

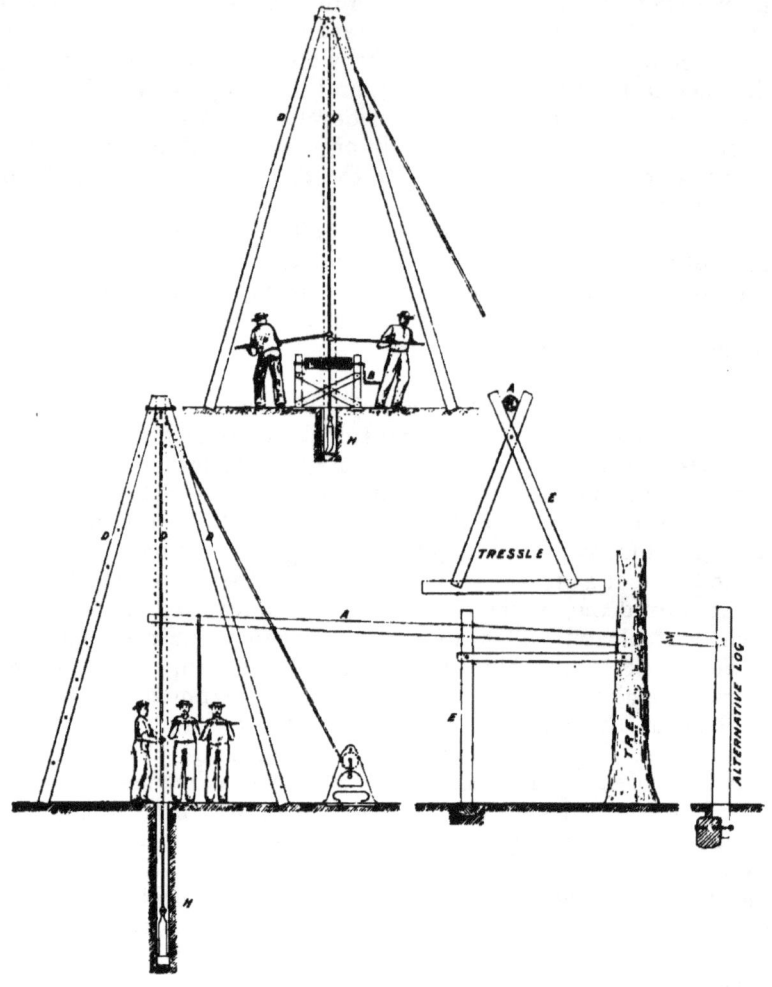

They are too complicated to manage for ordinary bush residents; liable to breakage, in which case repairs are difficult to make unless by a skilled mechanic; require either steam or horse-power to work them; and their cost is also much greater than

Sub-Artesian or Shallow-Boring. 123

necessity requires to perform the work. Let us, therefore, see what can be accomplished by that extremely simple, but quick and powerful system, the improved "Spring-pole" plant, putting aside all complicated apparatus with its greater cost, greater weight, risks of breakages, and in many cases incessant repairs.

The accompanying illustrations of this spring-pole plant show a side view and an end view of the derrick D, the borer to start the boring and the rock drill at work. The derrick consists of three poles of bush timber, 20ft. long, 5in. butt, with a bolt through at the top, to which are slung a large and a small pulley, working at right angles to each other for the drill and the sand pump ropes, respectively, to work through from a large (C) and a small winch (B) at the foot. The drilling is done by means of a spring-pole (A) 25ft. long, 5in. butt, preferably of ironbark, lancewood, or any elastic hardwood. The butt end of the pole is tenoned and inserted at 10ft. from the ground into a hole to fit it cut in a tree, or failing that, into a vertical log secured to a cross ground log. At 7ft. from the butt end the pole rests upon a trestle (E), also of bush timber, as shown in the diagram, the height of the tressle being 10ft., which throws the working end of the pole higher than the butt. A short hanging rope, with a cross piece of wood forming two handles for two workmen to pull down upon is attached to the pole at the working end. Another short rope, also attached to the same end of the pole over the bore-hole, has an iron rope clamp at the end of it level with the cross piece. In drilling in rock, after the cutting chisel or bit has been lowered to the bottom of the bore-hole by the large winch, the end of the pole is pulled down, the drill rope is then clamped on to the short rope connected with the end of the pole, the end of the pole let go, when its back spring lifts the drill, the movement being kept up by the two workmen pulling down, with little exertion, the end of the pole and again releasing it without letting go the handles.

Provision is made under this system by which any requisite power of spring, and lift and drop, may be obtained, and the force, speed and number of blows per minute (regulated to a nicety by the workmen) are greater than in any other movement —not excepting that of the deep artesian machinery—I am cognisant of or can conceive. After carefully noting the working of the various other machines in use, the movements for lifting and dropping drilling tools I have found this spring-pole movement gives from 40 to 50 per cent. more blows per minute than any other, and it is, therefore, the most economical in practice, as it is by far the simplest in construction and working. The spring of the pole being direct from the bearing on the tressle (E) is extremely sensitive, and the blows can, as pointed out, be adjusted in every way with the greatest nicety and precision.

In rock drilling (as I before remarked, I am not now treating of deep artesian drilling) with other machines the power has to pass from the shaft through wheels, levers, pulleys, &c., before it operates in lifting and dropping the chisel and in "tightening up" or "recovery" after the blow has been delivered in order to give another one. The pole, on the other hand, lifts and drops by its own action and spring, direct within itself; the power applied being that of two workmen in starting and keeping up with moderate exertion the movement, no steam or horse power being required.

An important feature of this apparatus is that, from its extreme simplicity, the timber work—*i.e.*, the derrick, springpole, tressle, and the sand-pump winch—can all be made in the country of bush timber (the iron work only for these being supplied), the carriage and liability to breakage in transit being thus saved and avoided. The weight of the winch and all iron work, including boring rods for 30 ft., complete set of chisels and other tools, is $1\frac{1}{2}$ tons.

Ordinary wrought-iron boring rods are used, if necessary, for commencing the bore in alluvial ground for not more than 30 ft., or until hard ground is met with, when the drill is used, worked by the spring pole. The rods are turned by hand spanners, and are lowered and raised by the larger winch, the rope from which passes over the larger pulley at the top of the derrick.

A "sand pump," consisting of a 5 in. tube, with a valve at the bottom for drawing up the cut rock and cleaning out the borehole, is worked from the small winch. The large winch is an ordinary quick-movement Contractors' winch with a break attached, for quickly lowering and raising the drilling bit. The small or sand-pump winch is made of bush timber, the iron work only being supplied. The working tools for rock drilling are *identical in every particular* with those used in deep artesian drilling, excepting that they are lighter. They consist of a chisel or bit, rimer, sinker bar, jars, eye-piece for drill rope attachment, undercutting bit, sand pump, ground-clamp and clamp spanner for lifting, lowering and turning the tubing, casing cap for driving the tubing, &c. The tubing used is the ordinary swelled-joint 6in. artesian, the cost of which at the bore would be about 4s. per foot. This tubing is perfectly watertight, and may be obtained "slotted," that is with slots

Sub-Artesian or Shallow-boring. 125

cut in the bottom tubes to admit water freely in coarse water-bearing strata.

In large sections of Australia, particularly of country east of the Ranges in which artesian water is not likely to be found, but where shallow sub-artesian water has been proved by a great number of successful borings to exist, the larger artesian machinery is not, on account of its greater cost and expense in working, admissible. Where it is used in procuring the invaluable deeper water the upper sub-artesian water found in most of the bores is unavoidably shut off by the tubing, which it is necessary to use in order to reach the larger supply below.

In order to confirm the position I take in advocating the importance of utilising the sub-artesian supplies, I may complete this section by giving an account of a sub-artesian formation which exists in many localities in the Western States of America,* and has been found to be of great value. the equivalent of which has been found in Australia, notably in Queensland by the Government Geologist (Mr. R. L. Jack) :—

"The term grit is descriptive of this formation everywhere, yet it is of varying constitution. In places it encloses a fine powder, but the powder is largely siliceous, and appears to be volcanic in origin. Elsewhere the grit is an aggregation of sand and lime, which we call its mortar-like form. Again the lime exceeds the sand in quantity, and it is sufficiently fine to be used for inside plastering. Then we have the mortar form, enclosing abundant pebbles, quartz, feldspar, diorite, greenstone (hornblende), and more rarely granite with other igneous rocks. Then the limy matrix almost disappears, and we have a heavy conglomerate of water-worn pebbles of the rocks above mentioned, with jasper, quartzite, and agate from the size of a nut to that of a large apple. The mortar-like form often hardens into a building stone, and its softer beds contain hard, tough nodules like indurated—not silicified—chalk. The conglomerate form changes at times into beds of a fair quality of sandstone. In some places it shows as a hardened bed of gravel with well-marked cross-bedding.

"This description of the Tertiary grit is that of a formation by no means confined to Kansas. While it is found in at least 20,000 square miles of that State. it is also conspicuous in an equal area of Colorado, a larger area of Western Nebraska, and in similar areas of New Mexico and Texas. Its character for holding water is continuous through all that region. On the level

---

* Report of Professor Robert Hay, F.G.S.A., to Senate. August, 1890.

prairie it is covered up by what we have called Tertiary marl, a loess-like deposit not altogether impervious to water, but so much so in most of its area to hold down the accumulated waters in the grit. The marl varies from 20 or 30 to 200 feet in thickness, and the grit from 20 to 100 feet. Through all this region the high prairie wells are sunk through the marl, and find abundance of water in the grit. Neither steam nor wind pumps reduce their supply. The conditions in Meade County, Kansas, which result in an artesian supply of water are strictly local, but it is very probable that there may be many repetitions of these local conditions in this vast region.

"Further, all the numerous springs of the region are from this grit, and many of them may be tapped above their present outlet and their waters made available for extensive irrigation. All the rivers of the plains, as distinguished from those whose sources are in the mountains, have their origin in the superabounding waters of the Tertiary grit. Many of these rivers are merely storm-used arroyos till they have cut below the grit, and then their channels have perennial streams, except where hidden in sand, which is largely the *débris* of the grit.

"The sandy nature of formations, which in a large part of the plains region are noted for their water-bearing capacity, is the main cause of the conditions which allow some of the river valleys to have a *sub-flow of water* equal to, or perhaps greater than, that of the visible streams. The conditions of hydrostatic pressure under which the sub-flows exist suggest that their phenomena are directly related to those of artesian wells and springs, and may properly be investigated with them."

The above description, although emanating from a distant portion of the crust of this earth, can be readily imagined to apply to many undiscovered and untried portions of this country. It will, it is hoped, afford further encouragement in our efforts to develop similar sub-artesian water-bearing strata in Australia.

Divining Rod.

Hydro-geology, or the "divining rod," has long been used by experts to designate the exact location of water veins below the earth's surface. Numerous wells, located by its professors, have proved successful; many have proved failures. Hydro-geology, as it is termed, is the use of the hazel or peach tree rod (or, failing these, it must be *willow*) to point out the presence of water-courses in the earth. It has been practised for a long period of time, and has of late years been elevated in France to the dignity of a science. The mode of operation is as follows:— A natural fork or twig is taken, having its limbs of equal size and length, and the leaves stripped off to the main stem. The ends of the limb are grasped firmly, one in each hand, with the back of the hands downwards, being at the same time extended from the body. The assertion of believers in the science is that the existence of water beneath the surface will be indicated by the revolving of the forked end towards the earth, having the two extremities for an axis. The explanation given by its experts is that there is a kind of magnetism by which the rod is disturbed and set in motion, and that the possession of the same is limited to *certain persons*. This explanation is given for what it is worth; one's inability must, however, be confessed to see any philosophical reasoning in it. Such persons as those alluded to have had fair and elaborate trials in Australia during many years, some of them in my presence, but the fact of there being at this present time no recognised practitioners who have followed up the business and adhered to it tends to show that the confidence of landholders in their favour has not been gained.

Section XXIV.

COST OF BORES.

In the last report of the Water Conservation Department of New South Wales, the following prices are given as paid for the earliest artesian bores made in that colony :—

Surface to 1000 ft. ...	27/- per foot.	
1000 ft. to 1500 ft. ...	35/-	,,
1500 ft. to 2000 ft. ...	40/-	,, exclusive of tubing.

the average cost being about 35/- per foot. The average cost of the early Queensland Government bores was about 37/- per foot.

"When the outlay necessary"—says the report—"to place a plant upon the ground, that must in the first place be incurred, the cost of the plant—probably more than £2000—before any return can be made, and the risks of the work which fall upon the Contractor, and the isolated and outlying portions of most of the bores are taken into consideration, these rates appear to be reasonable." Many of the risks attendant upon the early contracts have now been modified. In districts which have been tested the risks are reduced, and this with more systematic and business-like methods in carrying out the whole operation is considerably reducing the cost.

The best mode of executing future contracts for boring appears to me to be by means of the operations of a strong company, by which contracts could be carried out in a systematic and business-like manner at a fair price, and by which, through its financial strength, easy terms of payment could be arranged in order to meet lessees of grazing farms, as well as others who are under the present system more or less precluded from obtaining artesian water.

The cost of the Barcaldine bore (Government), Queensland, was 25/- per foot, the Contractors being supplied with wood and water, and free rail and carriage for all material. To-day bores can be made for from 8/- to 15/- per foot, the Contractors finding everything excepting tubing and wood and water.

CONCLUSION.

THIS book has been completed after recent visits to New South Wales, Victoria and South Australia, and it is hoped that enough information has been given to enable those who have neither the leisure nor inclination to study technical documents or Government Reports on the subject to form a general knowledge of the prominent features and conditions of artesian supplies, and the most important conclusions drawn will doubtless be as follows :—

1. That there is an immense accumulation of water lying in the crust of this portion of our earth which, as it is drawn upon for our requirements, will, in all probability, be amply replenished in the future as it has been in the past.
2. That the exposed areas of the outcrops or intakes of the water-bearing rocks are very much greater than was at first supposed, and that an additional supply—previously unthought of—is provided by the rivers and creeks which cross those outcrops, and which must enormously increase the general subterranean supply, bringing water as they do from large areas of country independent of that confined to the outcrop alone of the water-bearing rocks.
3. That the quantity of water in the gravels of which the channels of our rivers consist must be very great, and as much, if not more, water passes through them into the gravels of the Cretaceous underground rivers, and finally into the vast porous beds of the Cretaceous formation — from which our deep supplies are being obtained—as passes into all the outcrop areas of that supply.
4. That the average rainfall over the interior of this country— at the lowest estimate some fifteen inches, of which less than two inches passes into the rivers, and even less than that reaches the sea through the only river-way existing (the Murray), leaves the remainder, after allowing for evaporation—say ten inches—to travel from its intake areas underground, filling basin-like receptacles formed of the water-bearing rocks, which becoming surcharged, the overflow wends its way through continuous porous strata, surcharging other like receptacles at a lower level, and finally reaches and discharges itself in the bed of the ocean.

5. That the general fall of the water-bearing rocks is so *gradual, and the velocity of the flow of water so slow, and its volume, spread over enormous areas, is so great, that the supply may be considered inexhaustible.*
6. That the outflow from the bores in the artesian basins in gradually reducing the accumulated water produces a current and draught on the fresher water of the intake areas.

I may also say, in conclusion, that I have endeavoured to treat this great and vital subject of subterranean water supply from the standpoint of a Civil Engineer, whose education, studies and practice embrace all knowledge bearing upon it. In doing this I may also say that I have had in view the obligation I am under, as an old member of the Institution of Civil Engineers, London, to get at the truth in my profession, and to promulgate it to the best of my ability. The coadjutors of that profession, the Explorers, have achieved great results in adding to our geographical and physiographical knowledge, but the remarkable fact is apparent that whilst their researches located good pastoral lands, timber, minerals, &c., nothing, so far as my reading goes, is discoverable in their reports and writings indicating the existence of artesian water bearing-rocks. The flora and fauna are carefully described, and specimens have been collected at what must have been, under some conditions, at the expenditure of inconceivable trouble and labour. At the same time no explorers of any country in the world could have had the need of water more closely brought home to them than those of Australia during their heroic exposure to the fearful conditions of the barren drought-devoured plains of the interior. It seems reasonable, therefore, that in future explorations an artesian water supply expert should be attached to the expedition.

In the first edition of this book (Brisbane, 1883) was the following, which I venture to again place on record:—" The artesian system of well-sinking has proved a great boon to the civilised world. After its established success through the wonderful results given by the wells of Paris some 50 years ago, it has been almost universally used. The formation of the crust of the earth and its general physical conditions being nearly the same in all countries admitted of this. The science of geology, initiated and developed in Europe, has been adopted with little or no alteration of its rules in Asia, America, and Australia. The rain falls from the clouds on to the higher lands percolates through the pervious strata, finds the lowest level possible in its passage to the ocean, into which the surplus water discharges itself, forming on its way underground receptacles or reservoirs, the bed of which consists of the compact impervious strata of the rocks. This artesian water may derive its source on higher lands many hundreds of miles distant, finding its way by underground conduits to the site of the boring. As in the wells of the Great Desert of Sahara, before mentioned, a sandy, parched and barren plain may cover at no great distance below a subterranean underflow of water ready at the will of the explorer to burst forth and change the aspect of the surface above from a condition of sterility and death to one of fertility and life."

Australia is, without doubt, destined to take the leading position in the art which has formed the subject of this work. The peculiar physical character of the country, as shown by the scarcity of surface rivers, and want of a constant and reliable surface supply of water must necessarily command this. Although endowed with unbounded physical wealth and resources and an incomparable climate, artificial water supply must be called upon to fill the void of its one great drawback—periodical droughts—be it by means of irrigation from rivers or from its now great established source of supply, artesian wells.

Whatever conclusions may be reached as to the need or otherwise of legislation upon the subject, this may fairly be said: That the facts presented fully justify present expenditures, and demand, because of their weight and importance, a full and serious consideration of the grave legislative, economic, hydrologic, and other physical problems involved. They vitally relate to the present and future administration of a large section of our public domain; they are intimately connected with the industrial security of a considerable and growing population; and they greatly concern the conservation and progress of our Western country.

Appendix A.

TABLE OF QUEENSLAND BORES.

GOVERNMENT BORES.

Name of Bore.	Depth. feet.	Galls. per diem.	Remarks.
Barcaldine	691	175,000	Artesian.
Saltern	978	17,200	,,
Blackall	1663	300,000	,,
Tambo	1002	200,000	,,
Cunnamulla	1402	540,000	,,
Charleville	1371	3,000,000	,,
Wellshot	1160		
Racecourse (east of Range)	1781	8,228	Artesian water; contains small quantities of carburetted hydrogen; unfit for domestic purposes; bore abandoned.
Stewart's	2000	—	No water; abandoned.
Laidley (east of Range)	2512	—	Small supply; artesian; water unfit for domestic purposes; bore abandoned.
Clermont	323	—	Abandoned by contractor.
Muckadilla	3262	23,000	Artesian.
Normanton	1500	—	In progress.
North Rockhampton	2046	—	Pump lost in bore; abandoned.
M'Kinlay	1002	350,000	Artesian.
Thargomindah	2650	670,000	,,
"65-miles"	2362	104,727	,,
Thompson Watershed	3319	20,000	,,
Winton	4010	1,250,000	,,
Prairie	1395	—	Sub-artesian; in progress.
Back Creek	180	72,000	Artesian; sunk by Railway Department.

PRIVATE BORES.

Aberfoyle, No. 1	496	—	Good supply struck in sandstone; volume and whether overflowing not stated.
,, 2	500	—	In progress.

Appendix A.—(Private Bores.)

Name of Bore.	Depth. feet.	Galls. per diem.	Remarks.
Afton Downs, No. 1	1496	330,000	Artesian.
,, ,, 2	844	500,000	,,
,, ,, 3	2871	300,000	,,
,, ,, 4	700	—	Good supply.
Albilbah, No. 1	1610	—	Small supply; sub-artesian; work stopped.
,, ,, 2	1030	—	Small supply; tools stuck in bore; water level 100ft. below surface.
Alice Downs	2143	144,000	Artesian.
Aramac Station, No. 1	650	2,000,000	,,
,, ,, 2	1011	1,750,000	,,
,, ,, 3	900	2,500,000	,,
,, ,, 4	729	1,500,000	,,
,, ,, 5	627	1,500,000	,,
,, ,, 6	260	—	Abandoned; no water.
,, ,, 7	830	1,500,000	Artesian.
,, ,, 8	300	—	Sub-artesian; water level 48ft. below surface.
,, ,, 9	1820	600,000	Artesian.
Avondale Grazing Farm near Cunnamulla	317	30,000	Sub-artesian; pumped by steam power; water level 35ft. below surface; water slightly brackish.
Ban-do	2090	2,000,000	Artesian.
Barcaldine Divisional B.	1362	1,000,000	,,
Barcaldine Meat Preserving Company	1789	800,000	,,
Barcaldine Downs, No. 1	2500	100,000	Artesian; boring being continued.
,, ,, ,, 2	2350	500,000	,, ,, ,,
,, ,, ,, 3	1852	200,000	,,
Barenya, No. 1	1467	200,000	,,
,, ,, 2	2290	200,000	,,
Beaconsfield, No. 1	2040	1,500,000	,,
,, ,, 2	2406	800,000	,,
Beaudesert	140	—	Good supply; sub-artesian; water level 8oft. below surface.
Birkhead	—	—	(7 bores) 90 to 200ft.; sub-artesian; good supply; pumped.
Boatman, No. 1	1500	4,200,000	Artesian.
,, ,, 2	1960	3,000,000	,,

Appendix A.—(Private Bores.)

Name of Bore.	Depth. feet.	Galls. per diem.	Remarks.
Bogunda Selection, Hughenden No. 1	895	4,000	Artesian.
,, ,, ,, 2	1260	—	Not determined; sub-artesian; water level 100ft. below surface.
Bowen Downs, No. 1	970	493,000	Artesian.
,, ,, 2	1374	1,500,000	,,
,, ,, 3	1112	864,000	,,
,, ,, 4	3109	60,000	,,
,, ,, 5	2400	600,000	,, bore being continued.
Brighton Downs, No. 1	2396	1,250,000	,,
,, ,, 2	919	—	In progress.
Brookwood, Muttaburra	3065	800,000	Artesian; completed.
Bunda-Bunda, No. 1	682	1,500,000	,,
,, ,, 2	Not stated	1,000,000	,,
,, ,, 3	,,	1,500,000	,,
Bundilla	1751	150,000	,,
Burenda, No. 1	2130	20,000	Sub-artesian; water level 34ft. below surface; water pumped by windmill.
,, ,, 2	—	200,000	Artesian.
,, ,, 3	1600	25,000	,,
Burleigh	416	70,000	,,
Burranbilla	1811	3,000,000	,,
Caiwarra, No. 1	1810	1,080	,,
,, ,, 2	768	1,000,000	,,
Caledonia	720	80,000	,,
Cambridge Downs, No. 1	542	700,000	,,
,, ,, ,, 2	367	400,000	,,
,, ,, ,, 3	616	230,000	,,
,, ,, ,, 4	597	600,000	,,
,, ,, ,, 5	540	300,000	,,
,, ,, ,, 6	599	200,000	,,
,, ,, ,, 7	841	1,500,000	,,
,, ,, ,, 8	849	500,000	,,
,, ,, ,, 9	624	900,000	,,
Carandotta	1006	—	In progress.
Cassillis	2300	40,000	,,
Charlotte Plains (Cunnamulla) No. 1	1662	567,000	Artesian.
,, ,, 2	1848	4,000,000	,,
Charlotte Plains (Richmond) No. 1	1300	—	Sub-artesian; water level 15ft. from surface; abandoned.
,, ,, 2	360	100,000	Artesian; quite cold.

Appendix A.—(Private Bores.) 135

Name of Bore.	Depth. feet.	Galls. per diem.	Remarks.
Charlotte Plains (Richmond) No. 3	2012	—	Sub-artesian; water level 30ft. from surface; just warm.
,, ,, 4	700	300,000	Artesian.
Charlotte Vale and Victo	1600	1,000,000	,,
Chatsworth	3130	—	In progress.
Cleveland Point	—	—	(2 bores) 100 to 140ft.; sub-artesian; good supply; pumped.
Claverton, No. 1	1753	3,000,000	Artesian.
,, ,, 2	1777	1,500,000	,,
,, ,, 3	1819	1,500,000	,,
Clover Hills Grazing Farm, Barcaldine	1300	30,000	,, bore being continued.
Coomburra	2030	1,500,000	,,
Coongoola, No. 1	1614	1,000,000	,,
,, ,, 2	1807	1,500,000	,,
,, ,, 3	1803	3,500,000	,,
Coreena, No. 1	350	—	No flow; abandoned.
,, ,, 2	904	1,500,000	Artesian.
,, ,, 3	1350	350,000	,,
,, ,, 4	1993	25,000	Sub-artesian; water level 70ft. below surface; supply pumped.
,, ,, 5	760	1,000,000	Artesian.
,, ,, 6	631	—	Sub-artesian; water level 5ft. below surface; supply pumped.
,, ,, 7	170	—	Sub-artesian; water level 28ft. below surface; supply pumped.
Compton Downs, No. 1	850	100,000	Artesian.
,, ,, 2	500	250,000	,,
,, ,, 3	550	500,000	,,
,, ,, 4	634	600,000	,,
Cronin's Ag. F., Barcaldine	1002	1,000,000	Artesian.
Currawinya	264	80,000	,,
Dagworth	3335	1,250,000	,,
Dalgonally, No. 1	100	—	,,
,, ,, 2	336	—	,,
,, ,, 3	292	—	,,
,, ,, 4	300	—	,,
,, ,, 5	300	—	Sub-artesian; water level 6ft. below surface.

Appendix A.—(Private Bores.)

Name of Bore.	Depth. feet.	Galls. per diem.	Remarks.
Dalgonally, No. 6	140	not stated.	Artesian.
Dalgonally Meat Works, Normanton	100	—	Salt water; abandoned.
Dalzell's, Coreena	1175	600,000	Artesian.
Darr River Downs, No. 1	2007	30,000	Sub-artesian; water level 70ft. below surface; supply pumped.
,, ,, ,, 2	1002	40,000	Sub-artesian; supply pumped
,, ,, ,, 3	3630	500,000	Artesian.
,, ,, ,, 4	1900	4,800	In progress.
Delta, Barcaldine	1340	500,000	Artesian.
Dillalah, No. 1	1947	3,000,000	,,
,, ,, 2	2900	1,700,000	,,
,, ,, 3	1993	2,500,000	..
Durham Downs	1900	—	small supply; in progress.
Dynevor Downs (Bingera Station)	148	450,000	Artesian. Several other bores from 110 to 201 feet deep, and with a flow of 40,000 to 70,000 gallons per diem.
Eight Mile Plains	110	—	Sub-artesian; good supply; pumped.
Eulolo, No. 1	1426	1,000,000	Artesian.
,, ,, 2	1220	400,000	,,
Evora	2398	420,000	,,
Fernlee	2112	4,000,000	,,
Fort Constantine, No. 1	665	250,000	,,
,, ,, 2	900	—	In progress.
,, ,, 3	744	500,000	Artesian.
,, ,, 4	150	—	In primary formation.
Fraser and M'Lachlin's, Barcaldine	700	180,000	Artesian.
Glenormiston	200	—	In progress.
Greenmount	—	—	(3 bores) 90 to 130ft.; sub-artesian; good supply; pumped.
Harmian Park	1700	3,000,000	
Hamilton Downs	3100	1,000,000	!
Home Creek	1760	175,000	Artesian; boring being continued.
Hughenden (M. Council)	1500	—	,, in progress.
Jacondal	1015	500,000	,,
Jondaryan	—	—	(5 bores) 80 to 130ft; sub-artesian; good supply; pumped.

Appendix A.—(Private Bores.) 137

Name of Bore.	Depth. feet.	Galls. per di.m.	Remarks.
Katandra, No. 1	2250	—	In progress; water level 40ft from surface.
,, ,, 2	1160	—	In progress.
Kilcummin	1822	—	Abandoned.
Lammermoor	800	—	Sub-artesian; water level 6oft. below surface; abandoned.
Landsborough Downs	2063	not stated	Sub-artesian; water level 10ft. below surface; supply pumped.
Lansdowne, No. 1	2540	68,000	Artesian.
,, ,, 2	3000	150,000	,,
,, ,, 3	—	—	In progress.
Leichhardt Grazing Farm No. 1	1820	900,000	Artesian.
,, ,, ,, 2	2488	500,000	,,
,, ,, ,, 3	1467	900,000	,,
Llanrheidol	2441	1,250,000	,,
Malvern Hills	3948	—	In progress; water level 58ft. below surface.
Manfred Downs, No. 1	177	22,000	Artesian.
,, ,, 2	128	12,000	,,
,, ,, 3	86	14,000	,,
,, ,, 4	210	10,000	,,
,, ,, 5	200	20,000	Sub-artesian; supply pumped
,, ,, 6	98	16,000	Artesian.
,, ,, 7	678	50,000	,,
,, ,, 8	760	525,000	,,
,, ,, 9	707	200,000	,,
,, ,, 10	733	250,000	,,
,, ,, 11	1076	250,000	,,
,, ,, 12	890	200,000	,,
,, ,, 13	690	50,000	,,
Marathon, No. 1	1090	1,000,000	,,
,, ,, 2	1482	1,000,000	,,
,, ,, 3	1640	400,000	,,
,, ,, 4	714	800,000	,,
,, ,, 5	955	300,000	,,
Marchmont	1500	250,000	,, bore being deepened.
Marion Downs	1360	—	In progress.
Maxwelton, No. 1	1474	1,000,000	Artesian.
,, ,, 2	1371	670,000	,,
,, ,, 3	1590	1,200,000	,,
,, ,, 4	1434	900,000	,,
,, ,, 5	1200	1,500,000	,,
,, ,, 6	1000	—	In progress.

Name of Bore.	Depth. feet.	Galls. per diem.	Remarks.
Milo	2669	—	In progress.
Milton, Brisbane (east of Range)	600	—	Abandoned; no water struck
Mona	2350	2,000,000	Artesian; may be continued to 2500ft.
Mount Cornish	2300	576,000	,,
Mount Enniskillen	40	4,000	,,
Mount Esk	90	—	Sub-artesian; good supply; pumped.
Murra Murra	1891	4,000,000	,, work temporarily suspended.
Murweh, No. 1	1230	136,000	,,
,, ,, 2	1158	600,000	,,
,, ,, 3	1784	500,000	,,
,, ,, 4	1800	3,000,000	,,
Thomas Newton's, jun., Grazing Farm (Stainburn Downs Resump)	1100	2,000,000	,,
W. J. Newton's Grazing Farm	978	1,500,000	,,
Nine-mile, Barcaldine, and Aramac Road	630	150,000	,,
Newstead Farm, Ilfracombe	—	—	In progress.
Nive Downs, No. 1	2675	24,000	Sub-artesian; water level 60 ft. below surface; pumped by windmill.
,, ,, 2	3600	—	Sub-artesian; water level 60ft. from surface; in progress.
Noorama Resumption	1760	2,500,000	
Noorama Run, No. 1	1502	2,304,000	
,, ,, 2	1632	3,456,000	
Northampton	1803	1,500,000	
Oakley, No. 1	600	—	Sub-artesian; water level 80 feet below surface.
,, ,, 2	950	—	Abandoned.
Pawella (Aramac Resumption)	1226	200,000	Artesian.
Redcliffe (Hughenden)	3200	—	In progress.
Richardson's Farm (Barcaldine)	746	1,000,000	Artesian.
Richmond Downs, No. 1	699	800,000	,,
,, ,, ,, 2	480	500,000	,,
,, ,, ,, 3	1166	—	Sub-artesian; water rises to within 15ft. of surface; abandoned.

Appendix A.—(Private Bores.) 139

Name of Bore.			Depth. feet.	Galls. per diem.	Remarks.
Richmond Downs, No. 4			883	800,000	Artesian.
,,	,,	,, 5	590	800,000	,,
,,	,,	,, 6	857	200,000	,,
,,	,,	,, 7	829	—	Sub-artesian; water rises to within 90ft. of surface; abandoned.
,,	,,	,. 8	808	not stated	Sub-artesian; water rises to within 35ft. of surface; pumped.
,,	,,	,, 9	663	800,000	Artesian.
,,	,,	,, 10	560	1,000,000	,,
,,	,,	,, 11	856	—	Sub-artesian; water level 45 ft from surface; pumped.
,,	,,	,, 12	641	600,000	Artesian.
,,	,,	,, 13	800	700,000	,,
Rocklands, No. 1			... 420	—	Sub-artesian; abundant supply
,,	,, 2		... 300	—	,, ,,
,,	,, 3		... 300	—	,, ,,
,,	,, 4		... 400	—	,, ,,
,,	,, 5		... 300	—	,, ,,
,,	,, 6		... 350	—	,, ,,
,,	,, 7		... 850	—	,, water level 280 feet from surface.
Rockwood	...		... 2900	—	,, water level 50 feet from surface; in progress.
Saltern Creek, No. 1			... 1400	250,000	Artesian.
,,	,, 2		... 1884	250,000	,,
,,	,, 3		... 1970	690,000	,,
,,	,, 4		... 1594	780,000	,,
,,	,, 5		... 1998	500,000	,,
,,	,, 6		... 1873	700,000	,,
,,	,, 7		... 1863	430,000	,,
,,	,, 8		... 1268	750,000	,,
Saxby Downs, No. 1			... —	—	No information to hand.
,,	,, 2		... 543	1,000,000	Artesian.
Sesbania...	...		.. 2700	1,000,000	,,
Shakespere's	...		... 1040	500,000	,,
Stainburn	...		... —	2,000,000	,,
Strathfield, No. 1			... 857	450,000	,,
,,	,, 2		... 420	36,000	Sub-artesian; water level 50 ft. from surface; pumping.
,,	,, 3		... 820	35,000	Sub-artesian; water level 8ft. from surface; pumping.
Sylvania, No. 1 ...			... 1000	—	Artesian; in progress.
,, ,, 2 ...			... 700	500,000	,, ,,
Tambo, No. 1 ...			... 1910	60,000	,,

Appendix A.—(Private Bores.)

Name of Bore.	Depth. feet.	Galls. per diem.	Remarks.
Tambo, No. 2	2760	155,000	Artesian; in progress.
„ „ 3	2010	—	Sub-artesian; water level 36 feet below surface; suspended.
Tara, No. 1	2847	250,000	Artesian.
„ „ 2	2931	500,000	„
„ „ 3	2213	204,000	„
Tarbrax ...	1830	2,000,000	„
Telemon, No. 1	700	100,000	„
„ „ 2	301	—	Good supply; artesian; boring being continued.
„ „ 3	634	100,000	
Thurulgoona, No. 1	1290	80,000	Artesian.
„ „ 2	1682	576,000	„
„ „ 3	1630	200,000	„
„ „ 4	2037	20,000	„
„ „ 5	1862	313,000	„
„ „ 6	1612	180,000	„
„ „ 7	1672	50,000	„
„ „ 8	1685	100,000	„
„ „ 9	1785	3,500,000	„
„ „ 10	1600	44,000	„
„ „ 11	1710	3,000,000	„
Tinnenburra, No. 1	1254	800,000	„
„ „ 2	1160	4,000,000	„
„ „ 3	1280	300,000	„
„ „ 4	852	1,500,000	„
„ „ 5	1490	1,000,000	„
„ „ 6	1331	2,000,000	„
„ „ 7	1430	3,000,000	„
„ „ 8	1518	4,000,000	„
„ „ 9	1466	3,000,000	„
„ „ 10	740	—	In progress.
Tooleybuck, No. 1	917	40,000	Sub-artesian; water level 30 feet below surface.
„ „ 2	531	33,000	Artesian; struck granite.
„ „ 3	807	—	Good supply; sub-artesian; water level 50 feet below surface; struck granite.
„ „ 4	456	—	Sub-artesian; some water at 180 feet; struck granite.
„ „ 5	441	20,000	Artesian.
„ „ 6	860	—	Large supply; sub-artesian; water level 2 feet below surface; struck granite.
Toorak, No. 1	1285	1,000,000	Artesian.

Appendix A.—(Private Bores.) 141

Name of Bore.			Depth. feet.	Galls. per diem.	Remarks.
Toorak, No. 2		...	1550	800,000	Artesian.
Tulby, No. 1		...	1570	2,000,000	,,
,,	,, 2	...	1791	630,000	,,
Uanda, No. 1		...	335	200,000	,,
,,	,, 2	...	362	150,000	,,
,,	,, 3	...	248	80,000	,,
,,	,, 4	...	397	—	Sub-artesian; water rises to within 23 feet of surface.
,,	,, 5	...	280	30,000	Artesian.
,,	,, 6	...	296	—	Sub-artesian; water rises to within 35 feet of surface.
,,	,, 7	...	226	100,000	Artesian.
,,	,, 8	...	509	200,000	,,
,,	,, 9	...	518	200,000	,,
,,	,, 10	...	525	250,000	,,
,,	,, 11	...	606	30,000	,,
,,	,, 12	...	658	—	Sub artesian; water rises to within 30 feet of surface.
,,	,, 13	...	440	250,000	Artesian.
,,	,, 14	...	753	300,000	,,
,,	,, 15	...	367	20,000	,,
,,	,, 16	...	194	10,000	,,
,,	,, 17	...	302	17,000	,,
,,	,, 18	...	222	20,000	,,
,,	,, 19	...	301	70,000	,,
,,	,, 20	...	642	300,000	,,
,,	,, 21	...	310	60,000	,,
,,	,, 22	...	243	50,000	,,
,,	,, 23	...	260	20,000	,,
,,	,, 24	...	214	25,000	,,
,,	,, 25	...	217	20,000	,,
,,	,, 26	...	303	—	Sub-artesian; water level 36 feet below surface.
,,	,, 27	...	342	20,000	Artesian.
,,	,, 28	...	300	—	Sub-artesian; water level 12 feet below surface.
,,	,, 29	...	502	—	Very little water; abandoned
Vindex ...		...	—	—	In progress.
Warenda, No. 1		...	1075	50,000	Sub-artesian; water rises to within 70 feet of surface.
,,	,, 2	...	184	70,690	Artesian.
,,	,, 3	...	340	19,160	,,
,,	,, 4	...	511	140,000	,,
,,	,, 5		715	750,000	,,
,,	,, 6	...	559	1,015,600	,,
,,	,, 7	...	777	375,000	,,

Appendix A.—(*Private Bores.*)

Name of Bore.	Depth. feet.	Galls. per diem.	Remarks.
Warenda, No. 8	315	20,000	Sub-artesian; water rises to within 53 feet of surface.
,, ,, 9	210	13,000	Artesian.
,, ,, 10	653	185,100	,,
,, ,, 11	60	600	,,
,, ,, 12	75	5,000	Sub-artesian.
,, ,, 13	590	324,000	Artesian.
,, ,, 14	440	172,000	,,
,, ,, 15	363	144,000	,,
,, ,, 16	172	39,300	,,
,, ,, 17	176	—	Good supply; sub-artesian; water level 9 feet below surface; pumped.
,, ,, 18	192	1,800	Artesian.
,, ,, 19	780	777,600	,,
Weelamurra	1589	180,000	,,
Wellshot, No. 1	3500	216,000	,,
,, ,, 2	2576	650,000	,,
,, ,, 3	2500	90,000	,,
Westland, No. 1	2930	69,000	,,
,, ,, 2	3300	90,000	,, boring being continued
Whitula, No. 1	2138	—	No water; abandoned.
,, ,, 2	1469	—	Water level 70 feet below surface; abandoned.
Woolerina	2480	3,000,000	Artesian.

APPENDIX B.

Articles of Agreement made this
day of   in the year of our Lord one
thousand eight hundred and   Between

(hereinafter called the said Contractor) of the one part and
                          in the Colony of
                  (hereinafter called the owner) of
the other part. Witnesseth that he the said Contractor does
hereby for his part and his executor and administrator covenant
and agree with the said Owner that he the said Contractor will
do the necessary work in putting down a bore for water at any
spot on the   in the
                        to be fixed by the said
Owner at a remuneration after the rate of
per foot in depth for the first one thousand feet in depth from
the natural surface of the soil beyond which last mentioned
depth the said Contractor shall not be compellable to bore should
water be not then struck. Provided that should granite be
struck at any depth then the said Contractor shall be paid for
the depth to which the boring shall have then gone and shall
not be compellable to proceed further. Provided also that should
granite be struck or water reached at a depth of less than five
hundred feet from the natural surface the said Owner shall
nevertheless pay to the said Contractor the same amount as if
the boring had been continued to a depth of five hundred feet
from the natural surface so that in no case shall the said
Contractor receive a less sum than
And will put down all necessary casing for such bore. And will
do all such work well and faithfully and in a workmanlike
manner. And the said Owner doth hereby for himself his
executors and administrators covenant and agree with the said
Contractor his executors and administrators that he will pay for
such work at the rates aforesaid provided such payments to
be made to the said Contractor or his duly appointed manager
or to any person duly authorised in writing by the said
Contractor or by his duly authorised manager to receive the

Appendix B.

same and to be made as aforesaid from time to time as such work proceeds after the rate of twenty-five per cent. upon the actual amount due according to the rates aforesaid for the depth of boring from time to time done and any balance upon completion of said work. Provided that subject to the provisoes hereinbefore contained no progress payment shall be made until a depth of three hundred and thirty-three feet from the natural surface has been reached. And will find provide and convey to where the said work is being carried on all necessary casing for the execution of the said work and all wood coal fuel and water for the said work and will also convey or pay for the conveyance of all plant tools and implements of the said Contractor to the said spot.

And it is furthermore agreed between the said parties that should any dispute or difference as to such work arise the same shall be settled by arbitration.

In Witness whereof the parties hereto have hereunto set their hands and seals the day and year first hereinbefore mentioned.

Signed sealed and delivered
   by the said

Witness to the signature of said
   Contractor

Witness to the signature of said
   Owner

Appendix C.

ARTESIAN WELL BORING TOOLS AND PLANT COMPLETE.

Machinery for Rig.

1 crank shaft, $3\frac{7}{8}$ in. diameter, with wrought iron crank, pin, and collar; 1 draw spool, all wrought iron, and 1 sand line spool, with brakes; 2 belt tighteners, all wrought iron; 1 wrought iron driving-wheel, 6·0 in. diameter; 2 wrought iron draw-wheels. 4·6 in. diameter; 2 spool chains and swivels; 1 wrought iron con. rod and brasses; 1 cast iron spring pole-jacket; 1 saddle and stirrups for beam; 1 noddle pin and bolts; 1 slipper-out, complete (steel wheel); 1 set braces for slipper-out; 3 rope-sheaves (turned); 1 stirrup and king bolt for spring pole; 2 hydraulic jacks.

Derrick Tools.

2 knock wrenches; 2 catch wrenches; 1 pair heavy tool wrenches; 1 bit lever and chains; 1 extra large iron wrench board and stool; 1 pole-holder; 1 joint clamp.

Pole Fittings.

60 sets pole joints and centre straps; 2 pole swivels; 2 drill swivels; 1 drill chain and swivel; 1 cowsucker and shackles.

Drilling Tools.

1 sinker-bar, $2\frac{1}{2}$ in. diameter, 30 ft. long, joint in centre; 1 sinker-bar, 3 in. diameter, 30 ft. long, joint in centre; 1 sinker-bar, $3\frac{1}{2}$ in. diameter, 30 ft. long, joint in centre; 1 sinker-bar, 4 in. diameter, 30 ft. long, joint in centre; 1 pair drilling-jars, $1\frac{1}{2}$ in., legs all steel; 1 pair drilling-jars, $1\frac{1}{4}$ in., legs all steel; 1 pair drilling-jars, $1\frac{1}{8}$ in., legs all steel; 1 pair drilling-jars, 1 in., legs all steel; 2 dutchmen, 1 top and 1 bottom; 1 ball or improved under-runner for 6-in. and 8-in. casing; 2 6-in. bits; 2 8-in. bits; 2 10-in. bits; 2 5-in. bits.

Sand Pumps.

1 sand-pump, 4 in. diameter, brass valve and seat; 1 sand-pump, 5 in. diameter, brass valve and seat; 1 sand-pump, $6\frac{1}{2}$ in. diameter, brass valve and seat; 1 sand-pump, hanger, and chain.

Fishing Tools.

1 pair spring grabs; 1 spud, 8 ft. long; 21 two-legged socket; 21 one-legged socket; 1 half-turn hook.

Appendix C.

Casing Tools.

1 set casing dogs, 5 in., 6 in., and 8 in.; 3 sets clamp bolts, nuts, and washers ; 3 steel casing shoes, 5 in., 6 in., and 8 in.; 1 casing chain ring, and hook ; 51 pairs iron casing clamps and bolts.

Boring Tools.

1 surface auger, 12-in.; 100 ft. boring stems and connections ; 1 stone-hook ; 1 mud-pump, 10 in.; 1 boring chain ; 2000 ft. $2\frac{1}{4}$ in. best drill-poles.

Engine and boiler complete with all mountings; cylinder 10 in. x 12 in. (equal to 25 h.p.—American Catalogue). Engine fitted with pump and feed-heater, and cylinder lagged with non-conducting composition.

Boiler entirely of steel, and tested for a working pressure of 60 lb. per square inch. All pipes, valves, and connections fitted and supplied. and engine mounted on substantial bed, and governor specially designed for well-boring purposes fitted on slide valve jacket.

Wrought iron rig and derrick complete with all foundation timbers, bolts, &c., &c., and all bearing and holding-down bolts for saddle, crank, shaft, spools, tighteners, and snatch-post; also hardwood walkins, beam and spring pole, with all mountings fitted, the whole being erected in place before despatch.

1 improved new design automatic casing clamp, fitted with alternate dogs for 6-in. and 8-in. casing complete.

Tools, Stores, and Sundries for Artesian Boring Outfit.

2 engineer's hand hammers, 1 ratchet brace, 1 monkey wrench, 2 yards best insertion, 3 saw files, 1 blacksmith's anvil, 6 pairs blacksmith's tongs, 1 large fuller, 1 hot chisel, 1 set swedges $\frac{1}{2}$ in. to $1\frac{1}{2}$ in., 2 engineer's hand-chisels, 6 assorted drills, 6 assorted files, 6 assorted hammer-handles, 1 blacksmith's bellows, 1 Wilkinson's vyce 5 in., 1 large flatter, 1 forge complete 1 cold chisel, 2 handsaws, 1 cross-cut saw and handles, 1 draw knife, 2 gouges, 1 claw hammer, 1 wood rasp, 1 chalk line, 2 2-ft. four-fold rules, 1 carpenter's brace and bits, 25 ft. best $\frac{7}{16}$-in. chain, 2 treble iron blocks, 10-in., 2 single iron blocks, 6-in., 2 single iron blocks, 4-in., 2 dozen lampwicks, 1 adze and handle, 5 assorted chisels, 4 augers (assorted), 1 hatchet, 1 square, 1 66 ft. Chesterman's tape, 1 Rabone level, 3 steel sledge hammers, 25 ft. best $\frac{3}{8}$-in. chain, 4 lb. hemp packing, 3 hurricane lanterns, 1 12-in. engine room lamp and reflector. 2 dozen lamp chimneys, 70 ft. best 12-in. leather belting for draw belts, 45 ft. best 12 in. indiarubber for driving-belts, 2000 ft. Bullivant's steel-wire rope for sand line, 3 draw ropes, each 110 ft. long, complete, with thimbles spliced in ends, &c. (steel wire).

INDEX.

	PAGE.
Introduction	1
Antiquity of Boring	3
French Bores	8
Algerian Bores	9
English Bores	13
Deepest Bores in the World	15
American Bores	16
Victoria	26
South Australia	28
Western Australia	32
New South Wales	38
Queensland	51
Pioneer Bores of Australia	68
Definition of Artesian Water	70
The Hottest Regions on Earth	73
Artesian Bores as Affecting Climate	74
Geological Conditions Affecting the Flow of Artesian Water	75
Permanence of Supplies	81
Irrigation from Artesian Bores	86
Analysis of Water, Temperature, &c.	94
Machinery	98
Mechanical Power derivable from Artesian Bores, and Town Supplies	113
Tubing, and Durability of	116
Sub-artesian or Shallow Boring	119
Cost of Bores	128
Conclusion	129
Table of Bores in Queensland ... Appendix A	132
Form of Contract ... ,, B	143
List of Artesian Well-boring Tools and Plant ,, C	145

www.ingramcontent.com/pod-product-compliance
Lightning Source LLC
Chambersburg PA
CBHW031443160426
43195CB00010BB/828